Master

In-

Enterta

Master In-Car Entertainment

Vivian Capel

Newnes Technical Books

Hayden Book Company, Inc.

The Butterworth Group

United Kingdom
Butterworth & Co (Publishers) Ltd
London: 88 Kingsway, WC2B 6AB

Australia
Butterworths Pty Ltd
Sydney: 586 Pacific Highway, NSW 2067
Also at Melbourne, Brisbane, Adelaide and Perth

New Zealand
Butterworths of New Zealand Ltd
Wellington: 26-28 Waring Taylor Street, 1

South Africa
Butterworth & Co (South Africa) (Pty) Ltd
Durban: 152–154 Gale Street

USA and Canada
Hayden Book Company, Inc
50 Essex Street
Rochelle Park
New Jersey 07662

Library of Contress Catalog Card Number 77-15331

First published in 1977 by
Newnes Technical Books, a Butterworth imprint
and
Hayden Book Company, Inc

UK and all countries outside North America: ISBN 0 408 00286 7
USA and Canada: ISBN 0 8104 0844 9

Typeset by Butterworths Litho Preparation Department
Printed in England by Chapel River Press, Andover, Hants.

Preface

In-car entertainment is now enjoying a tremendous boom. Once considered a luxury or a means of being 'one up' on the neighbours it is now as commonplace as entertainment in the home. The frustrations of modern driving conditions have no doubt contributed more than a little to this situation.

For the owner-driver who is baffled by the array of equipment and needs guidance in making a choice, would like hints and tips on successful installation, and advice as to what to do when things go wrong, this book has been written. Technicalities have been reduced to the minimum so that only a rudimentary knowledge of car electrics and layout is needed.

For the writing of this book the author has drawn to some extent on his book *Audio on Wheels*, which is recommended for motor mechanics and fitters, radio engineers, professional installers and anyone requiring an 'in depth' treatment of the subject from basic principles upward.

Contents

1. The car-equipment scene

If you take a visit to your local car accessory store, likely as not it will be piled high with entertainment systems and the bits and pieces that go with them. There will be items of every conceivable sort, country of origin and price tag. In short, there is a boom in car entertainment.

The reason is not perhaps too difficult to see. For the majority of drivers, the daily haul into the city or the weekend run to the vacation spots has become more tiring and frustrating, with jams building up at the slightest provocation and high traffic density making almost every manoeuvre hazardous. Something soothing to ease the irritation of the moment, or perhaps to stimulate the tired driver heading for home after the hard day's slog, is not only beneficial to the nerves but no doubt contributes to safer driving.

So, we want to install a music system in our car. But first, back to the accessory shop and that huge pile of hardware. Just how does one pick out the best system? Even the buffs are often daunted by the choice, so the ordinary car-owner can be forgiven for being bewildered. Let's just take a look at what is on offer.

Radio

The mainstay of the car entertainment business is the radio. It is certainly the one that has been around the longest. Even for such a familiar, well known object, the range is considerable. There are single-band radios, and those with multiband coverage; some work on the amplitude modulated (a.m.) system, while others are frequency modulated (f.m.). Some have decoders which enable stereo broadcasts to be received, while the majority are mono only. Then there is the choice between the manually tuned units and those where the stations are selected by means of pushbuttons. Some have fancy tuning devices, such as those that automatically range over the tuning scale until a station strong enough to give a listenable programme is received. There it stops, and if you don't like what you hear, you press a button and the tuning moves on to the next one.

Then there is the matter of power output. This is the amount of electrical signal that the unit will deliver to the speaker. Until recent Federal laws were passed, this was very much jungle territory in the U.S.A. Every manufacturer had his own method of stating the output power, some (very few) giving what is known to technicians as the *r.m.s. power*, which is the mean power for accomplishing work in a circuit. It has been called the 'honest-to-goodness power'. Then there was the rating which quoted the power value at the peaks of the a.c. waves. This was about double the r.m.s. rating, but not particularly meaningful, though it looked good on paper. If you accept more distortion by driving the output stage harder you can get an even higher power figure, so many ratings were for unacceptably high distortion levels. Then there was the so-called 'music power', which depended on the ability of an amplifier to give higher powers than the normal rating for a fraction of a second, when a music transient was being reproduced. To complicate matters further there were combinations of all these, with r.m.s. music powers and peak music powers, all for different permitted distortion levels.

Well, it's good that in the U.S.A. this confusion is a thing of the past. Equipment must now be rated according to the r.m.s. wattage, and if any other is used it can only be specified as a secondary figure appearing in small print. We can then make a direct comparison between one unit and another as to its power output, without juggling with different rating systems. In the U.K. there are no such laws, and power ratings for car equipment differ according to the particular standard chosen by the manufacturer.

Every car radio needs an antenna, but with even this apparently simple and straightforward device there is a wide choice of types and shapes. There are the *whips*, the *rings* and the *rods*. There are rods that stay put and rods that you pull up and down, and those that rise and lower automatically. Which to choose is in this case governed by considerations of appearance and practicality, but a choice must still be made, and when confronted with a range of different models there can still be an element of uncertainty.

Tape players

Sooner or later comes the question: is a radio sufficient for my needs or do I want taped music? There is a greater variety of programme material with the radio, as in most areas a number of stations can be received, and of course the programme for each varies throughout the day. In addition we have the advantage of up-to-date news reports, weather and traffic information and the like. However, even with the variety of programmes available, music of one's choice rarely seems to be broadcast just when it is wanted, and this is especially the case with lovers of the classics. So, one has to listen without enjoyment, or switch off.

The alternative is the tape player. Here, a small collection of favourite tapes can be kept in the car, and one selected to suit exactly the mood of the moment. This advantage is undoubtedly one of the main reasons why tape players are rivalling the radio in popularity for mobile entertainment. Here again though, the prospective user is confronted with a choice: which tape format is the most suitable for music on the move?

The bother of threading a tape through an open-reel machine definitely makes this arrangement a non-starter for a car system. So, it must be one of the pre-packaged types, where the tape is conveniently fitted up in a container which can be popped into the player with one hand and played without more ado. There are two basic systems which satisfy this description, the *cassette* and the *cartridge*. With the cassette, there are two reels or hubs just as with an open-reel arrangement, and the tape passes from one to the other; then it is turned over to play the other tracks. In the cartridge, there is a single spool which accommodates an endless loop of tape, i.e. the end is spliced up to the start. Tape is pulled from the centre of the spool and taken up on the rim. There is no need to turn over the cartridge or rewind it, because when it comes to the end of the programme it just starts over again.

Choice of Facilities

However, with both these formats there have been quite a number of variants brought out and promoted by different manufacturing companies. In the cassette field there has been the RCA, R3M, Philips, and DC International types, while there have been cartridges under the names of Fidelipac, Conlay, Lear-Jet, MGM, and quite recently the Japanese Hipac. Fortunately, most of these seem to have dropped out of the running to leave a more or less clear field to the Lear-Jet cartridge and the European Philips Compact Cassette. This makes matters rather easier, but tape players of both types have a choice of features and refinements.

Some cassette players for example have the facility of automatically stopping and ejecting the cassette when it comes to its end. Some have both rewind and fast-forward controls, while others have only one of them, so that to rewind on a fast-forward only machine, you have to take the cassette out, turn it over and fast-forward it. Others have an automatic reverse function whereby the tape will start going the other way when it has reached the end, and the head will 'read' the other tracks. In this way it is not necessary to take the cassette out and turn it over. Most cassette machines replay only, but there are some models that also record. This is just the job for the executive who wants to dictate a memo driving in to the office – always assuming that he gets there in one piece!

In the case of the cartridge player, the variations are fewer because there is no need to rewind or fast-forward, nor to eject the cartridge at the end. Tracks are switched automatically in all models, and all have a manual override to change tracks or repeat the last one played. There is now though a choice of straight stereo or quadraphonic, of which more will be said later.

For those who want the advantages of having music of their choice always available, yet who also would like to hear the news broadcasts and other facilities offered by the radio, there are units of both cassette and cartridge format that include radio as well. These tend to be somewhat of a compromise, especially with the radio, because the size is designed to fit the standard car dash-board radio slot. Hence, the station scale is usually rather small and not so easy to read, and tuning is manual only, there being no room for pushbuttons. Also, the radio section has as a rule only one or two wavebands.

Even so, these units make a good compromise, and the best of both worlds are thus available in a convenient form. Separate units can of course be fitted, but take up more room, and there is the added complication of extra loudspeakers, or switching a single pair from one unit to the other.

Cassette-to-cartridge converter

There is another option for those who can't decide between the two tape formats, and that is the cassette-to-cartridge converter. This is a device, described in greater detail in Chapter 6, that enables a cassette to be played on a cartridge player. It is quite an ingenious piece of gadgetry, but again it is a compromise, and cassette performance is not as good as it would be on a proper cassette machine.

Price

So there we have the car entertainment scene. To choose the best arrangement for our purpose, we need to know a bit more about the way the equipment works, and about the respective snags and advantages. This is what we shall be covering in the next few chapters, and afterward go on to see how the system is fixed in the vehicle, and how to get the best out of it. Even the best equipment needs attention from time to time, and there may be troubles that have to be put right, so we shall also take a look at this aspect.

A thing which puzzles many people is the way prices vary from one model to another, even when the models seem similar and offer the same facilities. Are we really getting value for that extra money? Would the cheaper one do just as well or would we later regret buying it?

This sort of question is not easy to answer because there are so many factors that enter in to the pricing. One product, for example, may be imported from a country where labour costs are low and there may be a favourable rate of exchange between the two currencies, so it sells for a much lower price than a similar unit from another country where conditions are the opposite, or than a home-produced instrument. On the other hand, import tariffs can artificially inflate the price of certain goods shipped in from abroad. The only answer is to compare units only with others from the same country of manufacture. Then it can be seen whether they are generally higher or lower than goods from other countries, or from home.

Then of course there is the method of retailing. Some come through local dealers, while others may be obtainable from mail order advertisements, through wholesale factors or manufacturer's distributors. Discount stores usually offer the same product at a lower price than a specialist dealer.

Cheapest not Always Best Buy

So, economic reasons quite divorced from the quality of the equipment often make big differences to the price, and it obviously pays to take this into account when shopping around for the best value. It doesn't always pay, though, to go for the lowest price. Will spares and service information be readily available for those imported models? While some foreign firms have established service agents, others may not, and the importers may handle only a limited amount of components. With lesser known makes, too, the firm may fold up or the distributor cease dealing with their products. Then replacements become unobtainable.

Even when the same model is offered at different prices, it is not always the best thing to go for the cheapest. The specialist dealer, although perhaps dearer than other sources, can offer help and service if required, while after-sales-service is difficult to get from the discount house, and almost impossible from the mail order supplier. The local filling station and car repair shop can often supply equipment and fit it too, but again while the mechanic may make a good job of the installation, he is not an audio technician, so repairs to the inside would have to be farmed out.

Coming to the equipment itself, obviously it is possible to cut corners to make marginal savings, and many manufacturers do just that. There are so many ways that economies can be made to the detriment of either performance or durability, and unfortunately they cannot be seen from the outside. So you have to take a lot on trust. Usually though, if one manufacturer has a number of models at different prices, you can be sure of getting what you pay for. A dearer model will not just have an inflated price, but have genuine improvements over the cheaper ones.

2. Mono, stereo or quad?

While radios as a rule reproduce a mono signal through just one loudspeaker, some sport a stereo decoder which enables high quality stereo broadcasts to be received. Most of the tape players, whether cassette or cartridge, give stereo reproduction.

Stereo sound

Most people are familiar with stereo from the home hi-fi, but stereo in the automobile raises particular problems. To understand these, we will briefly review how the stereo effect is produced, and reproduced.

When we hear a sound coming from the left, the left ear picks it up at full strength. It also reaches the right ear, but owing to the masking effect of the head, the volume is less. There is also another more subtle effect, in that a time delay is introduced. The sound reaches the right ear fractionally later than it does the left. Sound is made up of a succession of compression and expansion waves in the air, and so the delay means that the waves in one ear do not coincide exactly with those in the other – the technicians say that they are *out-of-phase*. Waves producing high pitched sounds alternate more rapidly than those associated with bass tones, so there is a bigger phase difference between the ears when hearing high sounds than low. Sometimes the sound reaches the other ear a complete wave cycle later than it does the nearest one.

As the sound source moves from one side to the other, these volume and phase differences grow less, until it reaches the centre position, when there are no differences at all. From here to the opposite side, the differences increase, but the opposite way. All this information is fed to the brain by the respective aural nerves, and it not only analyses what sort of sound it is and its significance, but computes the precise direction that it came from. Thus if the sound signifies danger, one can immediately start running in the opposite direction! Obviously this is an extremely valuable facility.

In the quest for greater realism in reproduced sound, the object of stereo is to fool the brain into thinking that sounds are not coming from two loudspeakers, one on the left and one on the right, but from a host of intermediate points between them. Thus by using only two channels, a spatial effect can be obtained that might appear to need many more.

This is done by using two mics or sets of mics at the original recording session. Originally only two microphones were used, arranged at right-angles so that sounds coming from the left or right were picked up by the respective units, and those from intermediate points partially by both, but with volume and phase differences similar to those picked up by the ears. When fed to two loudspeakers, the ears of the listener received the same difference patterns, thus creating the stereo illusion.

Multimic Technique

While this method is still occasionally used, it is more common to string a whole set of microphones out among the players so that each one is close to a performer. This it is claimed gives a more immediate, less distant effect which seems to be preferred these days. To get the stereo effect though, the outputs from all these units are mixed in varying proportions down into two channels, so that volume differences exist similar to what would have been produced if only two mics had been employed. The snag with this system is that the phase differences are not obtained and so the brain is fed with only half the information it is accustomed to in order to give spatial identification to sound sources.

This explains the effect noticed by some listeners who have not heard a lot of modern stereo reproduction: they cannot easily identify location in the stereo image except for sounds coming from the extreme left or right. The brain needs to learn to locate sources from the reduced information provided. With two-mic stereo recordings, though, they fare much better.

It is obvious that to get a true stereo perspective it is essential for the sounds to arrive at the listener's ears in the same proportion as they did at the stereo mics at the recording. If the phase differences are absent in the recorded material, then it is even more important that the relative volume levels be accurate, as the brain is relying solely on these to judge the location.

Speaker positioning

The ideal way of ensuring this is to position the speakers an equal distance from the listener. If there should be any stereo phase-difference information in the material, then equidistant spacing will also make sure that this arrives in the correct relationship.

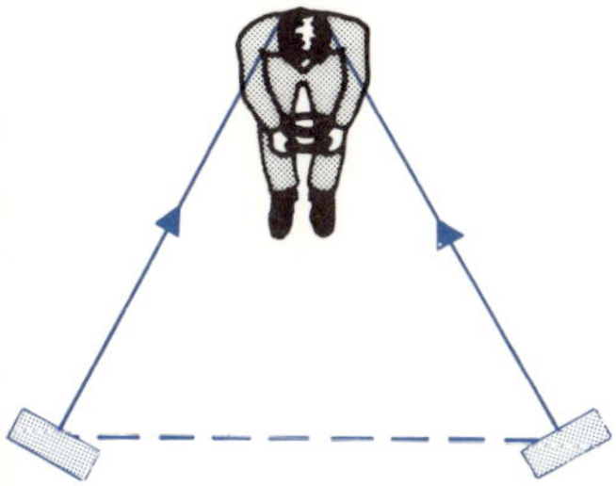

The ideal stereo configuration is when the two speakers and the listener are all equidistant. Sound from the speakers then arrives at the same time, and volume level is in correct proportion to establish an accurate stereo image between the speakers.

If the distances are not equal, but the difference is not too great, compensation can be made by adjustment of the balance control on the player. This increases the volume of one channel and decreases that of the other. The level of the further loudspeaker is thus increased, so that on a mono signal both appear to be of equal loudness.

The snag with this is that the sound from the nearer speaker arrives first, so a phase difference is introduced that was not part of the original sound pattern. If the distance difference is more than a few percent of the actual distance to the nearest speaker a phenomenon occurs known as the *Haas effect*. This makes it appear that all the sound is coming from the nearer loudspeaker; the furthest unit increases the loudness, but is not detectable as a separate sound source. If the distance is more than about 15 m (50 ft), then sound from the furthest speaker is heard as an echo, but this is not a situation likely to arise in either domestic or mobile stereo applications!

From this then it is clear that if the listener is much closer to one speaker than he is to the other, he will hear little from the furthest speaker unless the volume from it is much greater than that from the nearest, that is, when the stereo sound source is on the far side. Sources from about the middle to the near side will all appear to come from the near side.

Car stereo systems

With most car stereo systems, equal speaker distances are not possible, so the stereo effect is poor. Apart from extreme left or right signals, there is little difference between it and mono.

Where the speakers are mounted in or on the car dash-board, distances are unequal and also the angle is wrong. The stereo image is therefore poor. While some correction can be made with the balance control, this makes matters worse for a passenger.

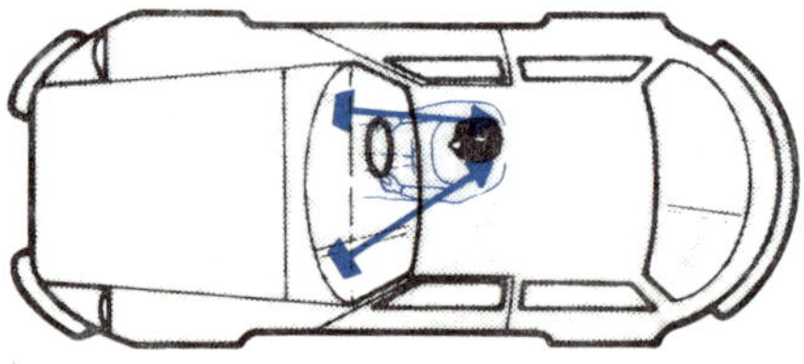

This arises from the usual positions adopted for the mounting of the loudspeakers. Some of these are: under the dash-board, one on the driver's and the other on the passenger's side; on top of the fascia, one each side; flush mounted in the left- and right-hand kick panels; and in the doors. In each one of these positions we have one speaker very close to the driver, and the other one at least twice the distance away on the passenger side.

Now while the driver can obtain some semblance of balance by using the balance control on the player, this will actually make matters worse for the passenger, as it is his nearest speaker that is now the loudest. But even if the balance control is so used, time lag will still upset the stereo positioning.

The situation is further aggravated by the fact that even where equal speaker distances are obtained, such as in a carefully set up domestic hi-fi installation, they should not be more than 60° apart relative to the listener. If this angle is exceeded, we get what has been described as a 'hole-in-the-middle' effect, where most of the sound appears to come from either the right or the left side and very little from in between. The stereo image is thus destroyed.

Mounting the speakers at the extremities of the front shelf extends the image further to the right, but the unequal spacing still remains.

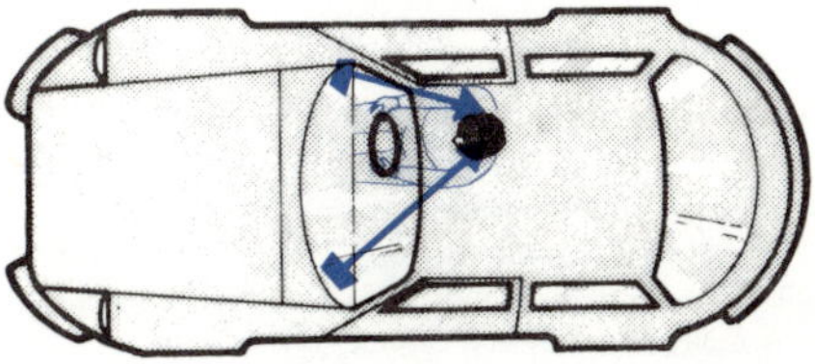

Further extension of the stereo image to the right can be made by mounting the speakers in the front door panels. Here, though, the space between the speakers and corresponding angles are such as to give a 'hole-in-the-middle' effect — all the sound appears to come as it actually does from either right or left and the illusion of centrally placed sounds is lost. Door position must be carefully chosen to avoid masking the speakers with the body.

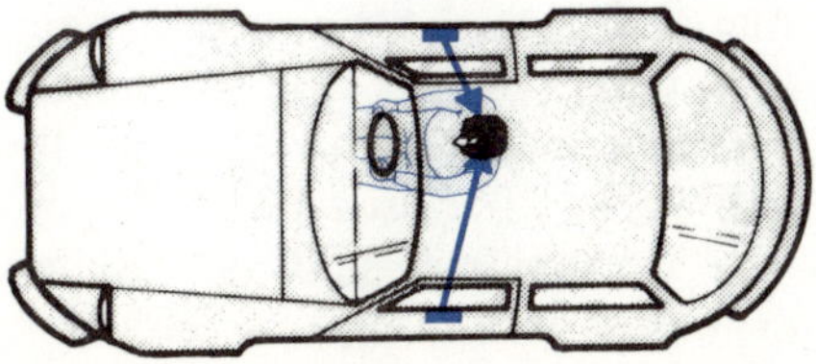

A moment's thought will make it obvious that in the case of many car installations, especially those where the speakers are mounted in the kick panels or in the doors, the angle formed between them and either listener is much greater than 60°.

So here is another reason why stereo is so poor in many mobile systems.

What we have to do then is to try to find some way of mounting the loudspeakers so that they will be of nearly equal distance between each one and the listener, whether driver or passenger. Furthermore, the angle between them and either listener should not exceed 60°. This seems a tall order, but it can be approximated, even though other snags may be introduced.

Rear Shelf Speaker Position

One popular speaker location is on the rear parcel shelf. Here the loudspeakers can be fitted in the shelf itself through holes cut for the purpose, or they can be surface mounted on the shelf. In this position the difference in distance is much less than with most conventional front positions, and also the angle can be made quite close to the ideal 60°.

Mounting the speakers on the rear parcel shelf equalises the distance between each one and the driver or passenger, so that reasonable stereo is possible. Obviously the image is placed unnaturally behind the listeners, and for rear passengers the stereo effect is non-existent, with one speaker right behind the head.

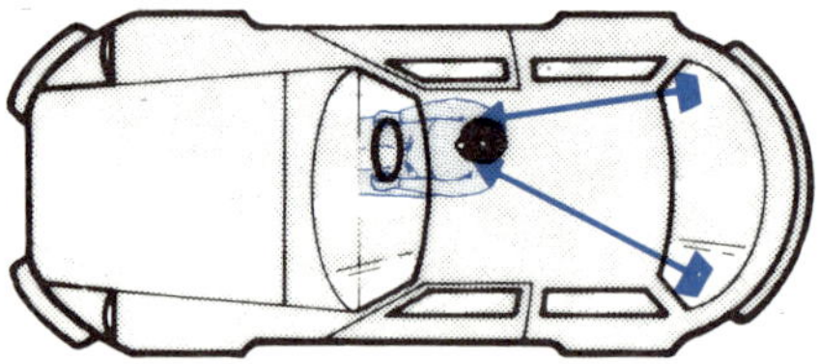

There are two main difficulties here, one being the rear stereo image. This is unnatural, as we do not normally sit with our backs to the performers. The other snag is that rear passengers get no stereo at all, except for anyone sitting in the middle of the back seat. With the normal side-seating position, each passenger will have a loudspeaker within a few inches of his head while the other will be several feet away. If rear passengers are seldom carried when the stereo is in use, then this drawback can be ignored.

While on the subject of rear parcel shelf speakers, it may be added that those fitted into the shelf will give a better bass response than the surface mounted variety. When the cone of a loudspeaker moves backwards and forwards it compresses the air on one side at the same time as it expands it on the other. If unmounted, the compression and rarefied regions would just merge at the rim of the speaker and little sound energy would be radiated (A). If the speaker is mounted on a flat board, usually called a *baffle*, the waves on each side are kept apart until they reach the edge of the board. The bigger the board, the longer the path between the sides and the greater the delay before merging and cancellation occurs (B).

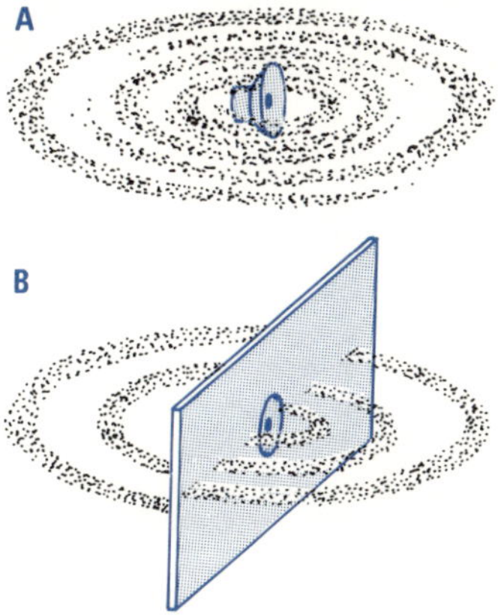

A Sound waves generated by the back of a loudspeaker cone are out-of-phase with those coming from the front; expansion of the air occurs at one side while compression takes place at the other.

B Mounting the speaker on a baffle keeps the two opposite sound patterns apart, and this reduces cancellation effects at bass frequencies, thus improving bass response. The larger the baffle the more bass is radiated.

Now the longer this path, the longer the wavelength that will be effectively radiated. As the long wavelengths of sound correspond to the lower notes, we thus get a better bass response. Speakers fitted into the shelf have a very long path because the front of the cone drives the air inside the car, while the rear affects the air in the luggage trunk.

Speakers mounted in small plastics cases have a much smaller path from front to back and so do not have the same depth of sound. This, incidentally, is why the kick panel and door mounted speakers have proved popular, because although the stereo image is very poor, there is a long front to back path and so good bass.

The fact that flush-mounted speakers on the rear shelf are facing upward is no problem if, as is usual, the rear window is raked over them. Glass is an excellent reflector of sound, so the sound from the speakers is thereby reflected forward to the front seats.

Windshield Reflection

There is another way of coming close to the ideal stereo position, and that is to mount the speakers on the top of the front fascia beneath the windshield, but facing forward towards the glass. The curvature and rake of the windshield is in most cases just right for reflecting the sound back in a way that gives an almost equal path from each speaker to both driver and passenger in the front, and

One of the best positions for good car stereo is with the speakers on the fascia top directed forward toward the windshield. Multiple reflections back toward the driver give some paths that are more or less equal. Glass is an ideal sound reflector and the angle and rake are about right. Both driver and passenger can enjoy good stereo.

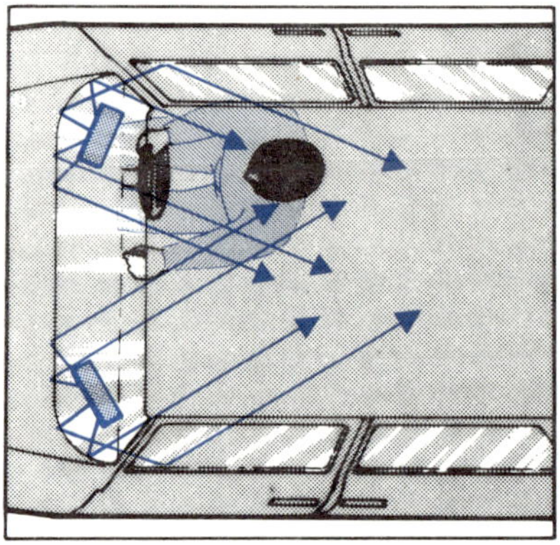

in addition to any passenger in the rear. There are multiple reflections from the windscreen, so the stereo image may be a little diffused, but it is a whole lot better than that produced by the conventional speaker placings.

The only difficulties are practical ones of actually fitting the speakers. Ideally, they should be flush-mounted in order to get a good bass response, but it is one thing to cut a large hole in a glass fibre rear parcel shelf or kick panel, and quite another to do so in a metal or padded metal surface of the fascia top. It may be wiser to opt for surface mounted units. These should be low so as not to interfere with the forward vision, but wide. Narrow elliptical speakers in suitable enclosures are therefore the best for this purpose.

Another possibility is to mount them overhead on the ceiling. Here again there are fitting problems – they cannot be screwed to the roof as the screws would come out on the outside. Magnetic feet is one possibility, but holes would have to be cut in the head-lining so that the feet would be in firm contact with the metalwork. There would also have to be sufficient headroom.

It can be seen then that there is no easy solution to the problem of getting a good stereo reproduction. Most installers tend to take the easy way out and fit the speakers in the most straightforward conventional position. However, by knowing the ideal requirements and the principles governing them, and exploring the possibilities in any particular model car, it is possible to get better results than are usually obtained.

Not everyone of course is particularly interested in stereo in the car. The equipment makers like to push stereo because it is dearer and they sell two speakers; in fact mono car players are few and far between. While stereo may be considered with some justification an essential ingredient of the home hi-fi system, is this necessarily true of the unit in the car? Are the objects in listening to music at home and on the road the same? Those who feel they are not and require only a background effect can opt for a mono outfit and save money as well as installation problems. Many users like to record some of their favourite music at home with a radio/cassette recorder, then play it back in the car. As these recordings are mono, a stereo outfit is not necessary to play them.

All this is a matter of personal preference, and many will insist on having stereo, even poor stereo, rather than mono, but the point is, you don't have to if you think it is not necessary. There are some neat combined radio/tape units in mono which are very convenient, combining versatility with comparative ease of installation.

Quadraphonic systems

Others may wish to go to the opposite extreme and choose a quadraphonic (quad) system. This comprises four channels and loudspeakers, and if properly installed adds another dimension of realism. Quad is used in two ways at present. With pop music the instrumentalists are scattered around the listener so that he is in effect in the middle, with sounds coming from all directions. It may add some interest, but can hardly be said to add realism, because no one hears music this way except those actually playing in the group. It is claimed though that the addition of directionality increases the artistic range and enables performers and recording director to get special effects not otherwise possible.

The more conventional use of quad is to add the reverberation which normally comes from all around one in the concert hall due to multiple reflections from walls and ceiling. This gives a sensation of actually being present to a far greater extent than ordinary stereo. So, you can enjoy a seat at the concert hall while driving home from the office. This is fine as long as you don't get carried away with the realism of it all and applaud at the end of the concert!

Although four loudspeakers must be installed for quad systems there is little problem accommodating the extra pair in the car. As these carry sounds coming from the rear, they can be mounted on or in the rear parcel shelf the same as with rear-mounted stereo speakers. If we are changing over from stereo and already have speakers in this position, they can obviously be used, but then we have to find room for the front pair.

With disc players there are problems in recording the four channels using a single groove with two walls. Several ways of doing this are in use. These can be divided into two main systems, the *discrete* type, where all four channels are kept separate by modulating two of them on to a carrier wave, the other two being recorded on the two groove walls like an ordinary stereo record, and the *matrix* type. This has the four signals mixed or encoded in such a way that they can be separated on replay. A matrix is used to do this but without other complicated circuitry the separation is imperfect. As good separation is necessary between the four channels, a discrete system is to be preferred if the advantages are not outweighed by problems in making the four separate recordings.

When using tape these problems disappear, as four independent tracks can be recorded. The cartridge is especially suitable for quad recording because it normally carries eight tracks to give four stereo pairs. These can be utilised to give instead two quad programmes, and in fact quad cartridges are available in this mode.

With cassettes matters are rather different. Although they have four tracks to give two stereo pairs, which theoretically could be used to give a single quad programme, in practice this would mean playing to the end of the tape and then rewinding, as there could be no 'return' programme. This would be rather inconvenient. Experimental cassettes have been made with eight tracks, but the tape is only half the width of cartridge tape, with extremely narrow tracks, and as background hiss and *drop-out* (momentary sound breaks due to loss of tape/head contact, uneven coating, etc.) increase with track narrowness, problems arise. Another experiment is the making of cassettes with matrix quad so that two programmes can be carried by the four cassette tracks. It seems rather a pity to have to do this to a medium that is inherently suited for discrete quad. No quad cassettes are commercially available at the time of writing, so the cartridge has the field to itself.

Surround sound from two channels

Quite impressive reverberation effects using four loudspeakers can be obtained from ordinary stereo tapes and stereo players, by using the Hafler technique. This is a simple method which relies on the fact that reverberation components are usually present to some extent in ordinary stereo recordings, although they are not separated from the main direct sound as they are in a proper quad recording. These reverberations travel a long path around the walls of the recording studio, being reflected from one surface to another many times until eventually they arrive at the microphone. Because of the random and varying lengths of these paths, reverberations picked up by the right microphone are completely out of phase with those arriving at the left. On the other hand, sounds coming directly from the performers, especially those situated near the centre of the stereo stage, are mostly in phase, but with minor differences between the two microphones.

If by some means we can separate signals that are similar but out of phase from those that are mostly in phase, we have succeeded in isolating the reverberations from the main body of direct sound, and they can then be fed to the rear loudspeakers.

The simple way of doing this is to connect the rear speakers in series with each other, across the positive loudspeaker terminals of the two stereo channels. The term 'positive' is used here not in the sense of a positive battery pole, because the amplifier output, being a.c., has no d.c. polarity. It denotes the loudspeaker terminal of each channel that is not connected to chassis or ground. A mono signal or centre stereo one will produce a wave of the same polarity at both positive terminals.

When now the signals in both channels are more-or-less in phase and equal in level, there is no voltage across the two positive terminals and so the rear loudspeakers produce no sound. When they are out of phase, opposite voltages are produced and the speakers respond. Thus they mainly reproduce the out-of-phase reverberation components, although they also will respond to in-phase signals of widely differing levels such as those coming from the extreme sides.

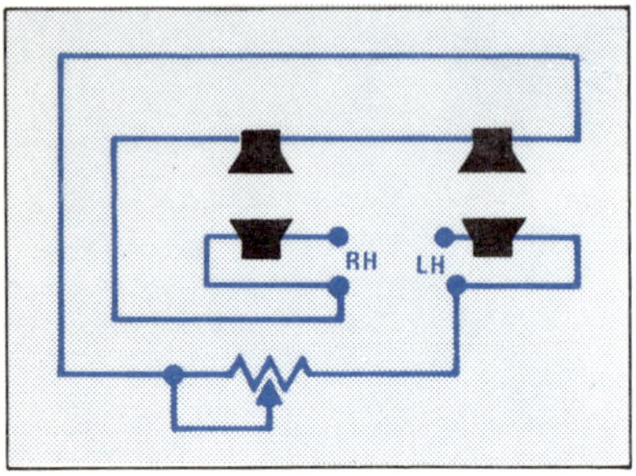

A 'surround stereo' effect can be obtained by connecting a pair of rear mounted speakers in series across the live terminals of the two stereo channels. This is known as a Hafler system. The rear speakers only reproduce out-of-phase signals, or those with wide level differences such as reverberation and echoes. Some crosstalk between the channels is inevitable though.

The success of this arrangement depends on the way the microphones were used at the original recording session. Where a large number were used close to the performers, the amount of reverberation is small and the effect is limited. If the two coincident stereo microphone technique was used, reverberation content is greater, and the results better. It can be expected then that some recordings will give a more impressive result than others. Unfortunately, most commercial ones now use the multimic technique.

Another factor is that both rear speakers, being connected in series, are producing the same sound. An improved effect can sometimes be obtained by running them out-of-phase with each other. Try reversing the connections to only one of the rear speakers, and see which connection gives the best effect. It should be mentioned that connecting the rear speakers across the front stereo channels in this manner results in crosstalk, i.e. sounds from the one channel will 'leak' through the rear speakers to the other. Some degrading of the front stereo image will result, but it is not too serious.

Some quad playing systems have a facility whereby ordinary stereo tapes can be played and given a *surround stereo* effect. When switched to this mode the rear speakers are fed with the out-of-phase signals as described previously, but they are supplied through separate amplifiers to prevent crosstalk taking place – hence there is no deterioration of the normal front stereo image. In addition, other phase shifts are introduced between the two rear speakers to enhance the random reverberation effect. Results can be quite impressive, but as with the simple speaker bridging circuit, it all depends on the way the original recording was made.

Two-speaker mono

Returning now to mono reproduction as obtained from a mono tape player or radio, it may if desired be fed into more than one loudspeaker. Usually, this means a speaker in the front and one at the back for rear seat passengers. There are a variety of ways in which the extra speaker can be controlled.

The simplest is by using a single-pole two-way switch (A). By this means either speaker can be selected, but not both at the same time.

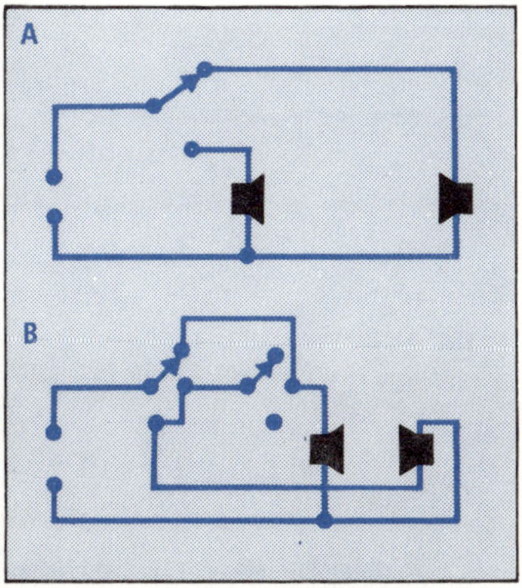

If two speakers are required (one in the rear) for a mono set-up such as an ordinary a.m. radio, either can be chosen, A, by wiring a simple two-way single pole switch. B If it is desired to select either back, front or both speakers, a two-pole three-way circuit must be wired as shown.

If the front one is required all the time but the rear one only occasionally, a simple on/off switch in series with the rear speaker will suffice. It can be either the toggle type, or rotary. If it is desired to select either speaker or both, a two-pole, three-way rotary switch will be needed (B). It can be wired for both speakers to come on in the middle position, or in the extreme clockwise position.

Volume control

The sensitivity of a loudspeaker varies according to the model, impedance and mounting method. When there are rear passengers, the rear loudspeaker will be situated just behind their heads, so the volume required will be less than that for the front speaker. On other occasions though, when there are no passengers in the rear, it may be desired to have the rear speaker volume at a higher level. For these reasons, some method of adjusting the relative volume levels is desirable. This can be done by means of volume controls wired into the loudspeaker leads.

To control the volume of an extra speaker, the circuit can be wired as shown. The wiper (moving contact) of the control is connected to one end and the control is wired in series with the extra speaker.

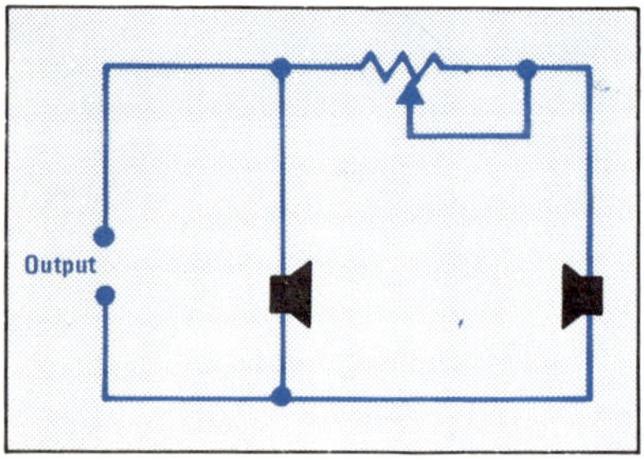

By using this circuit a volume control can be rigged to fade up one speaker and fade down the other at the same time. In the centre position volume should be about the same from both. There will be power loss with this circuit.

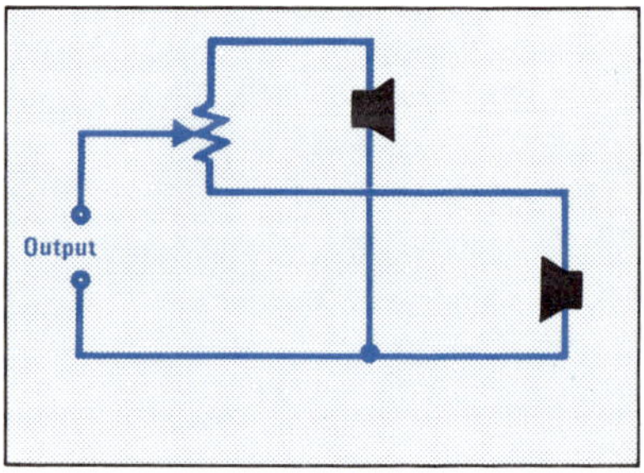

The easiest way is to connect a variable resistor in series with the rear speaker. This will reduce the volume relative to the front one, but will not make it greater. The control should be of the wire-wound variety and be rated at not less than 2 W. The resistance value is not critical, about 50Ω should do. Lower values will not give quite so much volume reduction in the lowest position, but will be smoother in operation.

Another method is to connect the speakers one to each end of the control, with the moving contact going to the radio output. When in the centre position, both speakers will give the same volume, but turning the control one way increases the volume to one speaker while decreasing it from the other, while turning it the other way does the reverse. A snag with this arrangement is that power is lost in the resistor at all settings of the control. To avoid too great a loss, the resistance must be kept fairly low, although this also restricts the amount of control possible between the two speakers. In the normal centre position, half the resistance will be in circuit with each loudspeaker. The proportion between this value and the speaker impedance represents the power loss in the circuit. The value then depends on the speaker impedance; for a 3Ω unit, the total resistance of the control should not exceed 15Ω, and the power rating not be less than 3 W.

In the case of a stereo system, for the sake of rear passengers it may be desired to use two extra speakers in the rear, fed with the normal stereo signal, not the out-of-phase reverberation components described earlier. The rear speakers can be connected in parallel across the front ones, but the circuit can be switched to give Hafler-effect reverberations when there are no rear passengers. This can be done by inserting a simple single-pole on/off switch between the centre connection of the two rear speakers, and the negative terminals of the stereo player which are joined together. When the switch is in the closed position the rear speakers carry the normal stereo signal, but when it is opened they function in the Hafler mode.

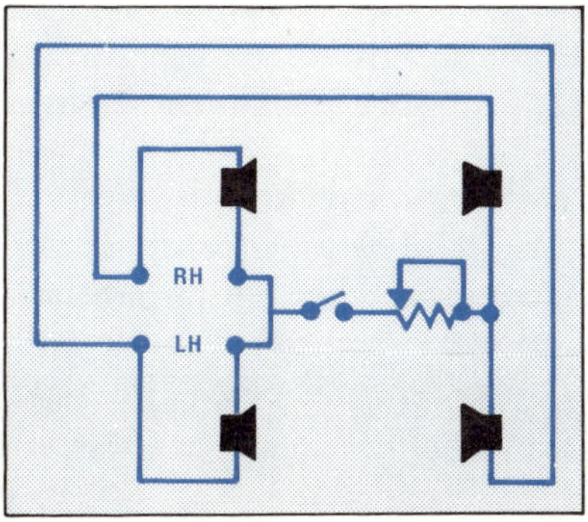

A Hafler surround stereo system with two extra speakers can be fitted with a switch so that the rear speakers can be switched to normal stereo if desired. The switch is wired in the common line to the centre tap.

If faders are required to control the level of the rear speakers, they can be connected in the same manner as for a mono single rear unit, but a pair of variable resistors ganged together with a common control spindle will be required.

A variety of switches and controls are made by many of the car-radio manufacturers, and examination of their catalogues will reveal a suitable component.

3. Mobile tape players

Without the invention of tape recording, recorded music in the car would be of a very limited nature. Special disc players were at one time made, but size was a problem, and it was only practicable to accommodate 7 in discs which had a limited running time. On long journeys it would therefore need many changes and a whole stack of discs to maintain music all the way.

Tape has solved the running time problem, as up to two hours can be accommodated on a single cassette, and recently even three-hour cassettes have been produced. So a handful of tape can now while away the longest and most tedious of journeys. Domestic tape recorders have been around for a long time now, but the mobile unit, although now well established, is a relative newcomer. There are certain necessary features in a car tape player which are not so important in a unit designed for use in the home, so we will take a look at these, but first of all some general points on tape recording principles.

Tape recording in general

The tape itself consists of a backing material made of some form of plastic, coated with iron-oxide particles held in a binding medium. Some manufacturers add traces of other materials such as cobalt to improve performance. Although one tape looks very much the same as another there are considerable differences in quality. The size and uniformity of the particles makes a big difference in performance, as does their distribution in the binder and the evenness

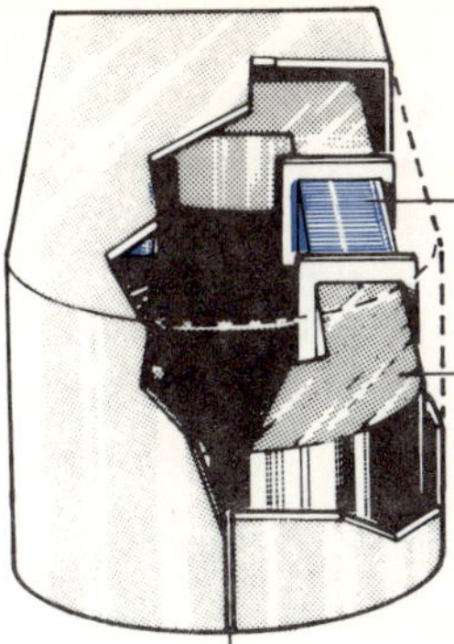

Construction of a record playback head in a tape recorder or playback head in a tape player. A magnetic gap is made in the iron core which is of laminated construction and on which the coil bobbins are fitted. Signal currents through the coils produce a corresponding magnetic field across the core gap. This gives rise to magnetic patterns in the tape which passes over it.

of the coating. Other factors involving the strength and elasticity of the backing, all play an essential part. It does not really pay, then, to make small savings on a cheap tape of unknown make, but on the other hand there is no point in paying the top price for the highest grade tape for mobile use where motor and road noise as well as outside distractions will cover any minor blemishes. This applies mainly to cassettes that you may record yourself at home; pre-recorded tapes are usually of high quality materials.

The oxide particles possess magnetic properties, that is they become magnetised when subjected to an external magnetic field. This occurs when they pass the record head, which is in effect an iron ring on which are wound coils of wire. At the point where the tape passes there is a minute gap in the ring, and when signal currents are passed through the coils, a magnetic field appears across the gap. The field is proportional to the current through the coils, and so also is the magnetic pattern recorded on the tape.

A low frequency wave produced by a low pitched sound thereby gives rise to alternate areas of high and low intensity magnetism along the tape, that are widely spaced. In contrast, a high frequency tone produces a closely spaced pattern. A limit to the minimum width of the magnetic pattern is obviously set by the width of the head gap – it cannot produce a narrower pattern than its own width any more than a thick pen nib can draw a thin line. There is therefore a limit to the highest frequency that can be recorded.

Tape speed

There is another factor influencing the high frequency response, and that is the speed of the tape past the head. The faster it goes, the more spread out will be the magnetic pattern and therefore the higher the frequency that can be recorded with a given head gap. The relationship between these can easily be shown. If we have a tape speed of 1 in/s, it follows that for a note of 1 kHz (1000 cycles per second), 1000 complete waves must be accommodated within the distance, so each will occupy a space of 0.001 in. In the case of a 10 kHz note (10 000 cycles per second), each wave will occupy only 0.0001 in, and this would be about the limit that could be recorded at this speed with present-day head gaps. To go any higher we have to decrease the gap width or increase the tape speed. If the speed is doubled to 2 in/s, the 10 kHz tone will produce patterns 0.0002 in apart, and in theory we could go up to 20 kHz, to take us to the limiting wavelength of 0.0001 in. In practice, other factors enter in at these high frequencies so that the relationship is not directly proportional – in other words we cannot get quite double the frequency by doubling the tape speed. The figures we have quoted are round figures for illustration, and are rather better than those currently obtainable.

The standard tape speeds used for domestic recorders are 4.76, 9.53 and 19.05 cm/s (1⅞, 3¾, and 7½ in/s). At one time only 19.05 cm/s (7½ in/s) was considered good enough for good quality recording of music; 9.53 cm/s (3¾ in/s) was capable of fair results where tape economy was required, while 4.76 cm/s (1⅞ in/s) was used for speech only, where maximum economy was the main consideration. Maximum upper frequency response was around 12 kHz for 19.05 cm/s, 8 kHz for 9.53 cm/s, and about 4 kHz for 4.76 cm/s. Since then there has been considerable development in head design and manufacture of recording tape, resulting in a considerable improvement in the response so that the slower speeds can give acceptable results. Now 10 kHz and over can be recorded at 4.76 cm/s. This is especially useful for mobile players, because the size of the tape pack can obviously be decreased at the slower speeds as less tape is needed, and size is a major factor with car players. The speeds used for cassettes and cartridges are 4.76 and 9.53 cm/s respectively.

Replay head

The replay head is similar to a record head, and in fact the same head is used for both operations in most tape recorders. The magnetised particles set up a magnetic flux across the gap in the iron ring, which generates a voltage in the windings. There are several requirements for good quality playback, one being that the

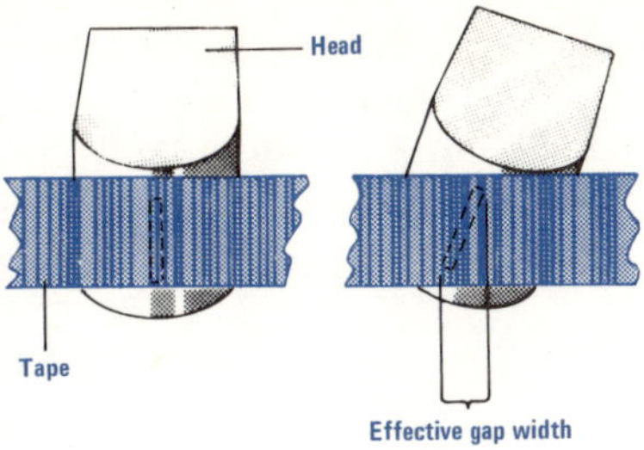

When a playback head is inclined so that the gap is not perfectly perpendicular relative to the direction of tape travel, the effective width of the gap is increased. It can therefore overlap several high frequency zones and the effect is to reduce the treble giving a muffled sound. The whole width of the tape is here shown recorded with a corresponding length gap; in practice the recorded track would be only part of the tape width.

playback head must have its gap aligned in the same plane as that of the record head. The standard alignment, or *azimuth setting* as it is called, is vertical so that the magnetic stripes are perpendicular to the edge of the tape. If the replay head should be tilted to one side or the other, then its gap will bridge across several adjacent stripes, resulting in a loss of high frequencies. It is as though the effective width of the gap had been increased. All recorders and players have an adjustment which enables the azimuth to be correctly set by rocking the head from side to side until a point is found where the high frequencies are at a maximum.

With car players, where there is a possibility of the setting altering with vibration or possible mishandling, this adjustment is more likely to need attention than with other machines.

Another important requirement is that the head be kept scrupulously clean. Any build-up of dirt on the face of the head prevents intimate contact between it and the tape, resulting in loss of volume and a high noise level. Even worse than ordinary dirt is loose oxide particles that may have been left by a tape. Being magnetic, these form a magnetic 'short circuit' across the gap. There is of course no physical gap, as this would soon attract the oxide particles and become filled in, so it is filled during manufacture with a solid non-magnetic shim so that the head face presents a smooth unbroken surface. In time the surface wears, and the shim may break, or the edges of the gap become uneven, thus leaving small crevices. In this condition the head may need frequent cleaning because the crevices soon fill with oxide and thus provide a key which encourages the formation of a large deposit.

Hand cleaning of the head with a swab and spirit solvent is the best, but with car players the head is buried inside the unit and not easily accessible without a lot of dismantling. A frequent run-through of a special cleaning tape is therefore the most practicable cleaning method. These tapes are made of a coarse weave fabric which should dislodge and remove small dirt deposits and polish the head off. They are not effective against heavy encrustations which have been allowed to build up, and so should be used regularly, before any need for cleaning becomes audibly apparent.

Car tape players

One big difference between car players and the domestic tape recorder is the absence of recording facilities in the former. This means that there are no recording bias or erose circuits, neither must the amplifier be switched from the record to play back mode, so the electronics are relatively simple. Another simplification is that as transistors run from low voltage supplies, the circuit can be fed directly from the battery, whereas a domestic recorder is fed from the 240 V house supply voltage circuit (110 V in the U.S.A.). This means extra circuitry involving a transformer and rectifiers.

Size is another factor. Some domestic instruments are quite bulky, especially reel-to-reel machines that accept the large 10 in spools. Even with the largest cars, space is not too plentiful around the dash-board, and with many of the smaller models, space limitations are greater.

Players must be designed to occupy the minimum of room, and both cassette and cartridge players are made with this in mind. Overall space in a car is at a premium, but in particular the room on the instrument panel. Tape cannot therefore be loaded with the spools or cassettes sideways to the operator as with the familiar home recorder, but must be 'mailed' through a slot in the control panel. Although more convenient to the user, this does introduce mechanical complications, especially with cassette players.

Controls must also be as simple as possible. It would be dangerous for a driver to carry out many control operations whilst driving. They are therefore reduced to the minimum. There is a volume and on/off switch, and many, though not all, have a tone control. The mechanism is started in most models by just inserting the cassette or cartridge, and some cassette players switch off and partially eject the cassette at the end of the tape. Cartridge players do not eject because they continue to play from the beginning, switching tracks automatically until the cartridge is pulled out from the unit.

It is the combined radio and player units that are rather more complicated, because of course there must be controls for both functions. Even here, though, the object of the design is to simplify matters, and in many models simply inserting the cassette or cartridge does the switching from radio to player mode, and removing or ejecting it switches back to radio again. Thus the only extra controls are for tuning and waveband.

Tape speed inaccuracies

A requirement of major importance with any tape player is to transport the tape past the head at an even and constant speed. Failure to do this results not merely in the music being reproduced at an incorrect speed, which for small variations is insignificant, but in changes of pitch which the ear readily detects.

Speed inaccuracies are of three main kinds. First there is the long term variation, involving a slow continuous deviation from the start to the finish of the tape. This is due to unequal loading on the transport system as tape is transferred from one reel to another and the speed of revolution of both reels changes. It is the least objectionable effect as the change is so gradual, and except for musicians who have the faculty of perfect pitch (and not even all musicians have this!) it usually passes unnoticed.

The remaining types of variation are termed *wow* and *flutter*, both of which are indeed noticeable. Wow is a regular cyclic speed change and is due to some rotating component meeting with unequal friction at different parts of its rotation, or to an eccentric rotating drive component. The frequency of the wow corresponds to that of the rotating component responsible for it, and can be up to ten variations per second. Faster variations are usually termed flutter and are not generally individually detectable. The effect is to impart a wavering or bubbling character to the sound, but in some cases it produces distortion not readily identifiable as speed variations, sounding more like an amplifier fault. Flutter is caused by the tape sticking at some point along its path and so proceeding in a succession of jerks.

It is impossible at this stage of tape transport development to get zero wow and flutter, but they can be reduced to a level where it cannot be detected by the ear. Thus there are wow and flutter specifications for all tape equipment, even that in the hi-fi class. Because of the working conditions with vibration and direction changes, automobile players are more vulnerable to speed variation than other types of tape unit.

Tape drive

The standard method of driving tape in most players is by means of a steel spindle against which the tape is pressed by a rubber pinch roller. This spindle is an extension of the shaft on which the flywheel runs, and it is the flywheel that is responsible for ironing out speed variations. The effectiveness of any flywheel depends on its speed of rotation and mass, particularly the mass at its periphery. The speed, and hence the efficiency, can be increased in two ways. Firstly, the tape drive spindle can be made of smaller diameter, so requiring a larger number of revolutions per second to achieve the standard linear tape speed. As the flywheel is directly coupled to the spindle it too will therefore revolve at a higher speed. The

A typical tape drive system. Tape is driven by a steel spindle against which it is pressed by a rubber pressure roller. The spindle is linked to a flywheel to smooth out small speed fluctuations, which is driven via a belt from the motor.

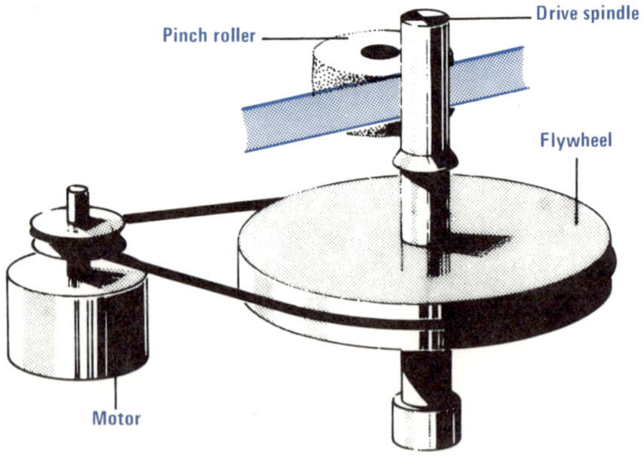

snag here is that by reducing the spindle diameter we also reduce the area of contact between it and the tape, thereby increasing the chance of slippage and reducing the traction. So speed variations from one cause are substituted for those of another.

The other method of increasing flywheel speed is simply to increase the linear speed of the tape. This is a further reason, in addition to improved frequency response, why higher tape speeds are used for the best quality reproduction.

Spindle diameter can be reduced if the tape width and size of the paying-out spool are also reduced, as the load pulled by the spindle is then less, and less traction is required. Thus cassette machines have small spindles and flywheels, but revolving at much higher speeds than cartridge players in spite of the slower linear tape speed.

While the flywheel plays an important part in keeping the tape speed constant, the motor must also run at a steady speed. With those run from the a.c. electricity supply this is no problem. The supply current alternates or changes polarity 50 times per second (60 times in the U.S.A.) and the motor is designed so that its speed is governed by this. In the case of the car d.c. battery supply there is no built-in speed synchronising feature, and speed must be controlled by other means.

Mechanical Speed Control

The most common method of doing this is contained within the motor itself and consists of an electromechanical governor. A short

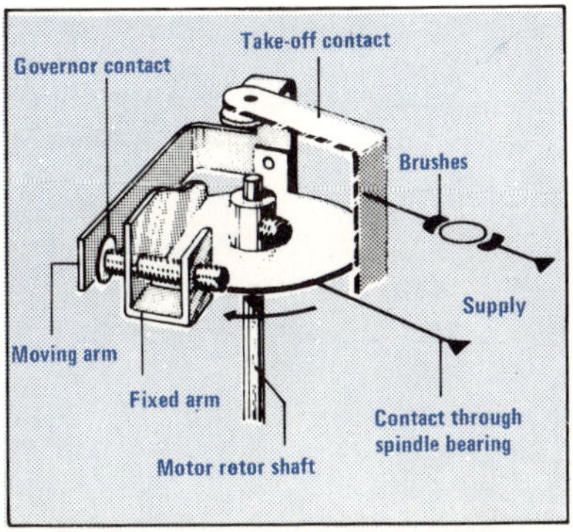

An electromechanical governor as mounted on a motor shaft. When the speed exceeds a certain point, the moving arm flies outward with centrifugal force, the contact opens and power to the motor is interrupted. When the motor slows the arm returns and the contact is re-made. Thus the speed is kept within close limits. The adjusting screw varies the point at which the arm loses contact and so can be used to set the motor speed.

springy metal strip is mounted on the motor rotor, having a contact and also a small weight fixed to its free end. The contact mates with another one fitted to part of the rotor assembly. Normally the contacts are closed, and as they are in series with the armature, current flows through them and the motor runs.

When reaching a certain speed the weight, attracted outward by centrifugal force, overcomes the tension of the strip and moves away, thus breaking the contact and interrupting the supply current. Immediately the motor starts to slow, causing the contacts to come together again and the motor speed to increase. This happens many times a second and the motor is thus kept at a predetermined average speed, although the instantaneous speed is continually oscillating above and below the average. The mass of the flywheel is too great to respond to these rapid changes and so effectively smooths them out to the average speed. If the flywheel is undersized though, as it often is with portable cassette recorders, it may be partially affected by the speed fluctuations, which is why the wow and flutter figures for these instruments are usually much higher than for the larger machines.

In practical governors, the weight is often dispensed with, as the mass of the moving contact serves the same purpose. Some means of adjusting the speed is necessary, and this is usually provided by a grub screw threaded into the rotor assembly. The tension between this and the contact strip and hence the mechanical resistance acting against the weight can thereby be varied. More often than not, the end of the grub screw itself serves as the fixed contact. If the tension is lessened, the weight moves earlier and the speed is reduced, but if increased, the speed also increases. The screw is accessible through a hole in the motor casing and should be adjusted by means of a timed length of tape or a stroboscope.

The operation of the governor entails the continual making and breaking of the contacts. This along with the motor commutator produces interference in the form of a background 'hash' that can get into the amplifying circuits. In fact it can be quite troublesome, needing special circuit filters and other measures to minimise it.

Electronic Control

Other problems associated with the electromechanical governor are: oxidisation of the contacts, pitting and wear, and displacement of the contact strip. The grub screw can also work loose if not sealed with a spot of paint after adjustment. For these reasons electronic control is being increasingly used. A fairly standard circuit has been evolved and is used by various makers, although minor details and component values are changed in different models.

For the sake of technical readers we will give a brief description of how it works, but at the same time we will try not to leave the non-technical ones completely in the dark. Two transistors are used for the circuit, the second being a power transistor supplying

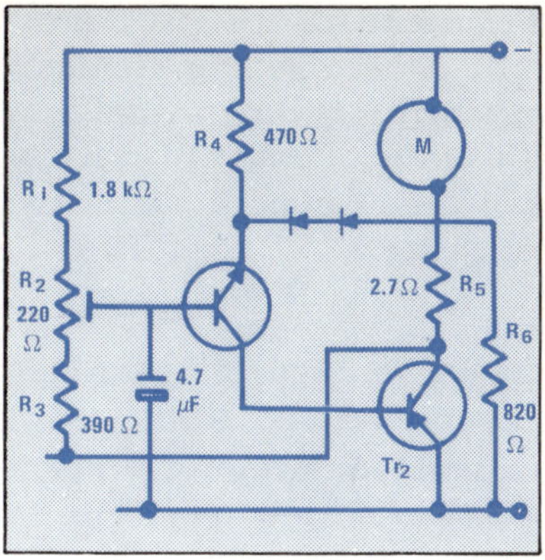

A simple electronic motor control circuit. Any fluctuation of the supply voltage or variation of current through the motor due to a speed change produces a change in voltage across R_5. This changes the bias to Tr_1 which controls the current through Tr_2, which in turn alters the current through the motor.

current to the motor. This transistor is biased so that it passes current, by transistor No. 1 (Tr_1), and this in turn receives its controlling current through a network of resistors (one of which is variable) connected across the motor.

If there is a drop in the supply voltage there is a fall in voltage across the motor, and this is communicated to the emitter circuit of Tr_1 through a couple of diodes. The voltage drop is also applied to the base of this transistor through the network of resistors, but because of their dividing action, the potential difference appearing at the base is less than that communicated to the emitter. There is therefore, a net *increase* in the voltage applied to the combined base/emitter circuit. This increases the current through this transistor, which in turn increases the forward bias to Tr_2, making it pass more current to the motor. Thus the drop in battery voltage is compensated for. Should the supply voltage rise, the reverse process will operate and the current will be reduced.

When a motor runs off load, it takes very little current from the supply, but as the load is increased it slows down and the current gets larger. The reason for this is that a motor also acts as a generator, and when running at full speed generates a current that is almost equal to the driving current, which being of opposite polarity nearly counteracts it. When it slows down, the reverse current is reduced and there is less cancellation of the original driving current.

This fact is made use of in the electronic control circuit to counteract any slowing of the motor due to extra friction or other factors. The increased current taken by the motor produces an extra voltage drop over the series resistor, R_5. As this component is also connected between the base and emitter circuits of Tr_1 there is an increase of voltage applied there. So the transistor conducts more, thereby causing Tr_2 also to pass increased current, and thus more power is applied to the motor. Therefore, if either the battery voltage fluctuates or the mechanical load on the motor varies, the circuit will regulate the power to compensate.

As we noted, one of the resistors in the network supplying the base of Tr_1 is a variable one. This makes a convenient control to adjust the motor speed. The control, being external to the motor, is easier to reset and less likely to be disturbed than the internal governor. The diodes are temperature conscious and so compensate for any temperature variation that may affect the operation of the transistors.

This simple circuit is quite effective, but there are more sophisticated arrangements for maintaining accurate motor speed. Another system uses a type of servo-control in which there is actually a small generator included in the motor. The output of this generator is fed to a circuit which is made up of a capacitor and resistor and is thereby frequency conscious. The voltage appearing at the junction between the capacitor and resistor depends on the frequency and therefore on the speed of the motor/generator. A diode rectifies this voltage and it is applied as forward bias to a two-stage transistor amplifier. The motor is controlled by the second stage as with the previous circuit, so the power supplied to the motor is controlled by the frequency, hence speed, of the motor.

The previous circuit only came into operation when there was a difference from the correct running speed. A small but definite time elapses before the circuit responds and compensates. Thus, like the electromechanical governor, there is a tendency for oscillation above and below the correct speed, which the flywheel must smooth out. With this servo-circuit, the control voltage derived from the generator frequency is always present and the slightest deviation produces almost immediate correction. Voltage and thermal stabilising are also included.

Motors

Motors that operate from the a.c. supply voltage have their rotors energised by a rotating magnetic field set up by the stator coils. As there is no varying voltage in the d.c. supply, d.c. motors need to have electrical contacts made to the rotor. This is done by a circular row of contacts called the *commutator*, and a pair of sliding electrodes termed *brushes*. Wear takes place between these sliding contacts, and interference is generated. In addition, the drag of the brushes on the commutator imposes a mechanical load which can never be perfectly even. So this type of motor has an inherent speed instability not found with a.c. motors.

To overcome this drawback some car equipment manufacturers use a special kind of brushless d.c. motor. In one model, three sets of motor coils and three secondary coils are wound on the stator assembly, the rotor consisting of a permanent magnet with a straight ferrite rod bisecting an aluminium disc fitted to one end. A transistor oscillator energises a small primary coil, and the flux is conveyed to whichever one of the three secondaries the ferrite rod happens to be aligned with. The induced voltage is

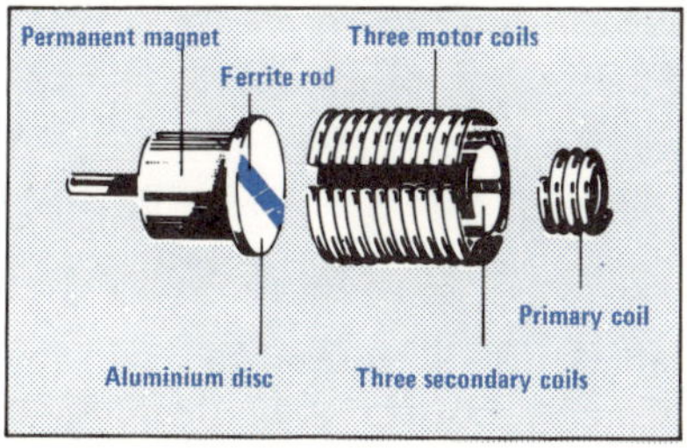

In this brushless motor, coupling between each of the three sets of primary and secondary coils is provided by the ferrite rod on the rotor. As each is coupled the magnetic field pulls the rotor around part of a turn, and this changes the position of the ferrite rod so that the next set of coils are coupled and so rotation is maintained.

applied to one of three switching transistors which thereby switches on power to one of the three motor coils. This attracts the permanent magnet of the rotor, causing it to rotate part of a turn. The ferrite rod also turns and now aligns itself to the next secondary coil. The next transistor is switched on by this and the next motor coil energised. Further rotation takes place and successive coils are switched into circuit, thus sustaining the rotation.

Speed control is achieved by sampling voltage pulses generated in the motor coils, dependent in value on the motor speed. These are smoothed and amplified, then applied to the oscillator as bias. As the strength of the oscillations determines the current induced in the secondary coils and these in turn control the current through the switching transistors and the motor coils, motor power is thereby regulated. A feature of this arrangement is that the frequency of the oscillator is not important and so does not have to be carefully controlled.

There are other types of servo system which use a synchronous motor similar to the a.c. electricity supply type, but they must be run from an oscillator that has the frequency controlled within very narrow limits. An a.c. voltage generated in the motor is compared with this in a discriminator circuit, and any difference is used to correct the oscillator frequency.

Take-up spool drive

Driving the tape past the head is not the only mechanical function that must be carried out where two spools are used to pay out and take up the tape. Tape is merely pulled off the paying-out spool by the tape drive spindle, which is straightforward enough; it is the take-up spool that introduces a complication. The problem is that the spool does not revolve at a constant speed, turning faster when empty at the start of the tape than when full at the end, and of course the speed reduction is continuous throughout the playing of the tape. The reason is that the circumference of the tape lying in the spool is constantly increasing as more tape is added, so each revolution picks up a greater length of tape. As the tape is being driven past the head at a constant speed, the spool revolutions must get progressively slower.

The solution to the problem is actually quite simple – to drive the spool through a friction clutch that is constantly slipping. A common form of clutch is a pair of metal or plastics discs with a

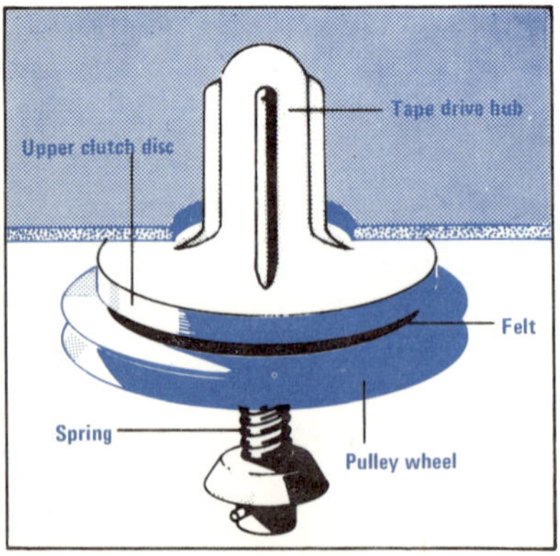

The take-up spool in a cassette player must be driven at an ever decreasing rate as the spool fills and the diameter increases. To do this a slipping clutch is included in the drive. This is a typical arrangement, with a felt pad separating the two plates, and pressure being maintained by a coil spring.

felt pad in between. Drive is applied to one disc and the spool is driven by the other. Slight pressure is maintained between them by a coil spring.

Fast wind

Another facility usually provided is the fast rewinding or winding-on of the tape. This is done by engaging a belt or idler wheel drive between the motor and the appropriate spool carrier. The normal tape drive system is disengaged during this operation, and the friction clutch bypassed. In some models the motor governor is shorted out to give maximum motor speed. Not all car players include rewind and fast-forward facilities, and some have just one of them.

4. The cartridge player

The most popular form of car tape system is the Lear-Jet cartridge. This consists of a single spool with a cheek on one side only, wound with standard 6.3 mm (¼ in) tape which is the same size as for open-reel machines. The ends of the tape are joined together by a short section of metallised foil tape to form an endless loop. Tape is pulled out from the centre of the spool by the tape drive spindle and pinch roller, and is fed back on to the outer layer and thus taken back up on the spool. The spool is not driven mechanically by the player, but turns only by reason of the tape being pulled from it.

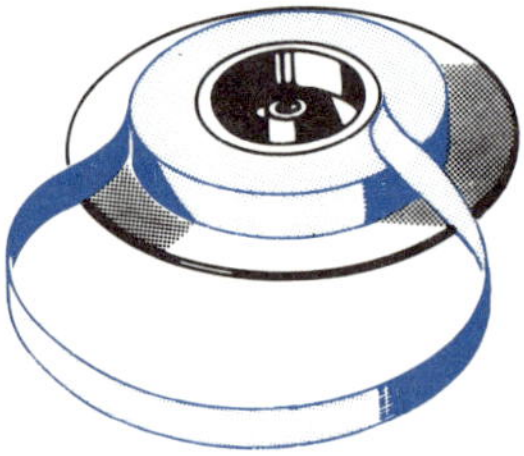

The endless tape as used in a cartridge. The spool has only one cheek at the underside, thus enabling tape to be pulled from the centre. As it does so, the spool turns and the used tape is taken up on the outer layer. Diameter difference between inner and outer layer would cause the spool to stall if special dry lubricant were not used in the tape coating to cause the layers to slip over each other. The ends are spliced together with metal foil which actuates the track changing mechanism.

At first glance this might seem an impossible arrangement. The circumference of the outer layer of tape is obviously greater than at the centre, so one revolution should take up more tape on the outside than could be pulled out from the inside. So it seems the action is doomed from the start and the spool must jam to a halt before a single revolution can be completed. This would in fact be the case if it were not for the special dry lubricant which is incorporated in the tape. Adjacent layers are thereby allowed to slip over each other, enabling more tape to be pulled from the centre than would normally be given up by a single revolution.

An even tension is maintained by this means, minimising the risk of the tape tangles sometimes encountered with cassettes. As the amount of tape on the spool is always constant, the mechanical load on the transport system is always the same, which is in contrast with cassette or open-reel systems where the load constantly changes as one spool empties and the other fills up. This ensures better long term speed constancy. There is, though, more wear on the tape than with other systems, and this is greatly accelerated when the lubricant wears off. Dirt, which is the enemy of all tape systems, is especially detrimental to the cartridge because it reduces the effectiveness of the lubricant.

From the spool centre, the tape passes over a guide and around a pulley wheel to some cut-out portions in the side of the cartridge casing. Behind the tape at this point is mounted a pad, usually of

Inside view of cartridge. Tape pulled from the centre passes over the right-hand guide, then along the aperture where the head makes contact with the rubber pinch roller on the left. Here the steel drive spindle grips the tape against the roller pulling it along its path. From here it is taken up on the outside of the spool. The spool itself is not mechanically driven, but turns by reason of the tape being pulled from it.

The disposition of the eight tracks on a 6.25 mm (¼ in) cartridge tape. All are interleaved so that the stereo channels consist of tracks 1 and 5, 2 and 6, 3 and 7, and 4 and 8. The playback head moves physically down the width of the tape so that its gaps successively read each pair. After the last pair it returns to the top to start again.

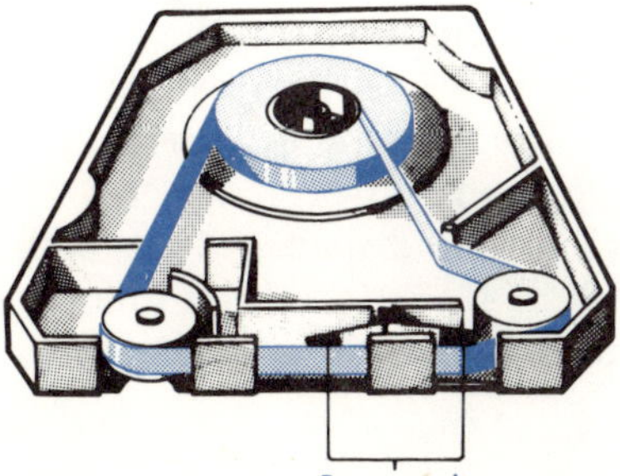

sponge plastic, which serves to press it against the playback head and thus ensure intimate contact with it. From there, the tape travels to the final cut-out where it passes around a rubber pinch roller. It is here that the drive spindle in the player engages the tape by pressing it against the pinch roller. After leaving the roller, it passes to the outside of the spool and the circuit has been completed.

Each cartridge, therefore, has its own pressure pad and pinch roller mounted internally.

The tape runs at 9.53 cm/s (3¾ in/s), the same as single-speed open-reel recorders. Whereas though there are usually two or four tracks recorded on open-reel tapes, the cartridge has eight. These are recorded all in the same direction in four stereo pairs. To avoid spacing the head gaps too close together, thus increasing the manufacturing difficulties, the stereo pairs are not consecutive but interleaved. So, numbering them from top to bottom of the tape, we have tracks 1 and 5 for channel one; tracks 2 and 6, channel two; tracks 3 and 7, channel three, and tracks 4 and 8, channel four.

Tape drive

Where two spools are used, as with open-reel and cassette systems, the mechanical drive arrangement is quite complicated, because there must be reduction drives and clutches to drive the take-up spool, and also fast-forward and rewind facilities. All these are eliminated with the cartridge player as there is no independent take-up spool, and rewinding is not necessary because the tape is automatically at the start of the programme when it is finished.

Some players include a fast-forward facility so that any desired portion of the tape can be quickly selected. But if you should overshoot the required part, there is no going back, and you have to continue to fast forward until the part comes round again. The fast forward speed is not all that fast compared to the normal running speed. This is because the tape has to be driven by the drive spindle and flywheel, there being no other mechanical connection to the spool, whereas with open-reel and cassette systems the take-up spool is driven directly and the tape disengaged from the spindle and pinch roller. Fast-forwarding, then, is only of marginal advantage and a machine should not be chosen on this feature alone.

The mechanical drive system of a cartridge player is therefore very simple. It consists basically of a flywheel driven by a belt from the motor, the tape drive spindle being an extension of the flywheel shaft as it is in most other types of tape machine. When the cartridge is pushed into place through the slot in the player, the tape is pressed against the head and the tape pinch roller engages against the drive spindle. As pressure must obviously be maintained, the cartridge case is retained in its working position by means of a roller mounted on a sprung arm engaging with a notch in the cartridge side. Some models also use a spring mounted long roller against either the top or bottom of the cartridge to prevent lateral movement of the unit in the playing slot.

With most models a switch is actuated by pushing the cartridge home into the player, and this switches on the motor and the amplifier circuits. Thus there is no need for a separate manual switch, and the controls are thereby simplified.

Track changing mechanism

As the tape continues to run indefinitely, the same channel would be played over and over again unless some means were included of changing over to the other tracks. This could be done by means of a manual control, but would need frequent attention, as the maximum length of each channel is about twenty minutes, and most are considerably shorter than this.

The changeover is accomplished by means of the metallised foil included between the end and beginning of the loop. The tape runs over a pair of electrical contacts, which when the foil appears are shorted out and an electric current initiates the track change cycle.

While details of the changing mechanism may vary from one model to another, most follow a basic pattern. The current passes through a solenoid, which is thereby energised, causing retraction of a

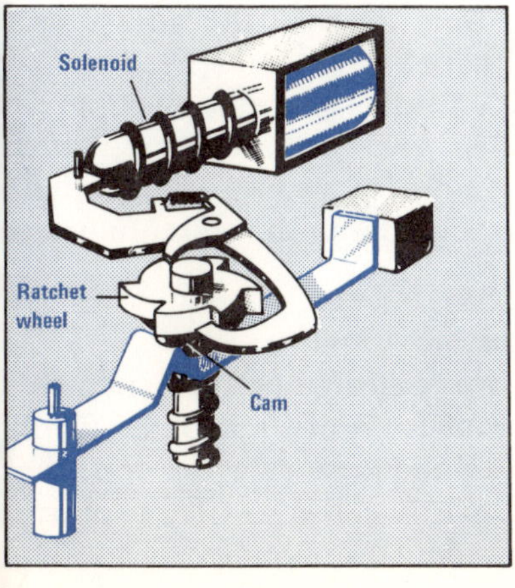

A typical track changing mechanism. The solenoid is actuated by electrical pulses supplied by the end-of-tape foil passing over contacts. This operates the ratchet, turning the toothed wheel a quarter of a turn. A cam on the lower part of the wheel progressively pushes the head carriage downward against the tension of a coil spring. After its lowest position the cam allows the spring to push it up to the top again.

spring-loaded plunger. This is coupled at its free end to an escapement mechanism consisting of a toothed ratchet and pawl. Retraction of the plunger causes the pawl to move the ratchet wheel a quarter-turn. The lower part of the ratchet wheel is formed into a cam with four steps and this engages against a spring-loaded bracket which carries the playback head. Thus at each quarter turn, the cam depresses the bracket downward another step and the head gaps line up opposite the next track. After the final and lowest set of tracks have been played, the cam allows the bracket to rise to the first position, and the series starts again.

Head gap adjustment and other controls

The head gaps must line up accurately with the recorded tracks, which in view of their narrow width calls for some precision in the changing mechanism. If the register is not perfect, part of the gap will overlap and will 'read' part of the margin in between the tracks, while part of the track will be missed on the other edge. The result is loss of volume and an increase in noise. If the error is greater, the gap will overlap the adjacent track and two tracks will be heard at low volume at the same time. Obviously some form of adjustment is needed to compensate for misplacement and wear in the system.

This is provided by means of an internal set-screw which alters the position of the head bracket relative to the cam, or the head relative to the bracket. In the latter case, the screw must not be mistaken for the head azimuth adjustment screw which rocks the head from one side to the other in order to get a perfectly vertical setting for the gap. With some models, the adjustment is brought out to the front control panel so that the head height setting can be altered if required without any dismantling. In such instruments the facility is termed *fine tuning control*, which can be rather confusing if the unit includes a radio. The control is adjusted to give maximum volume together with minimum noise and adjacent channel interference.

Another control sometimes found is a 'repeat' button. When this is pushed in it will remain in this position until released, and its function is to open-circuit the tape contacts so that the foil does not initiate a track change. Thus the same track will be played over and over until the button is released.

It is useful to know which track is being played, and most cartridge players have a track indicator. This is worked by means of an insulated disc linked to the ratchet wheel, which carries an electrical contact. As the wheel turns through the four quadrants the contact mates with one of four stationary pairs of contacts, shorting them out in turn. These are wired to four lamps mounted behind an indicator panel which thus show which particular track is being played.

In addition to the repeat button, a manual track change button is usually provided. It is connected across the tape contacts, and actuates the changing mechanism when pressed in the same way as does the foil.

The cartridge and quadraphony

As there are eight tracks on a cartridge, the format is ideal for quad. Quad, we remember from Chapter 2, needs four tracks which ideally should be separate from each other and so free from mutual crosstalk. Most of the controversy over rival disc systems is centred on the fact that this cannot be done perfectly with only two channels – the two groove walls. Here though we have the separate channels available, and quad cartridges give two programmes of four tracks each in place of the four stereo ones.

The player must have four amplifiers and of course four speakers, two mounted at the front and two at the rear. Just as volume and tone controls for stereo two-channel players are ganged so that a single control operates the elements in each amplifier at the same time, so those of the four amplifiers are ganged up to a single knob. An extra balance control is needed to give front/back balance in addition to the normal right/left balance.

The head has four magnetic gaps, arranged to read tracks 1, 3, 5, and 7 for the first programme, and tracks 2, 4, 6, and 8 for the

Track disposition with quad cartridge tape. Just two channels of quad are recorded using the eight tracks. These are interleaved so that the first channel consists of tracks 1, 3, 5, and 7, while the second uses tracks 2, 4, 6, and 8.

second. This rather complicates the track change. Going from programme 1 to 2 means just moving down a single space, the same as for stereo. In going from the second back to the first though, the head must jump three positions, that is gap No. 1 must go from track 2 to 3 to 4 to 1. This needs a special circuit, using another solenoid which produces three pulses to actuate the main changeover solenoid three times. It does this only on alternate switchings so that the head moves only one position when going from programme 1 to 2, but three positions when going from 2 to 1. This circuit is also controlled by the manual track change switch.

Stereo cartridges can also be played in quad players. The special track change circuit is switched out, and also the extra amplifiers, otherwise two different tracks would be heard from the rear speakers. In some models the amplifiers are left on but are switched in to the front ones. No switching is necessary from the user as this is done by a feeler in the player engaging in a channel cut in the top left-hand side of the cartridge casing.

5. The cassette player

Hitherto, the cartridge has been the predominant format for mobile player systems. The Philips cassette has been popular for some while for home audio systems and portable tape recorders, but it is only recently that it has shown any signs of being a serious contender to the cartridge in the field of car players.

Construction of the cassette

In place of the single spool of the cartridge, the cassette has two spools lying side by side, tape being pulled from one and fed to the other in the same manner as with open-reel recorders. As one of the requirements of a mobile system is compactness, this may seem to be a waste of space. In fact less space is used than would be required for an open-reel system of similar tape dimensions. The reason is that the spools have no side cheeks, being really hubs rather than spools, and so can be placed closer together than spools with cheeks. The tape on one hub when it is full extends more than halfway towards the other, but this causes no problem because as the second hub fills, the first one empties, so there is a constant gap maintained between them.

Tape is prevented from spilling by the sides of the cassette, which means that the inside dimension of the cassette must be only slightly greater than the tape width. Friction is minimised by special liners which reduce the drag on the tape edge as it revolves.

The tape passes from one hub over a guide post to a nylon roller and then travels down the inside of the long edge of the cassette to another roller at the opposite end. Along this side there are a number of cut-outs, and behind the middle one is mounted the pressure pad on a phosphor-bronze spring. This is the point where the playback head engages. There is another cut-out for the erase head in a cassette recorder, but this is not used in a player. Just behind the last cut-out a round hole is drilled right through the

Inside view of a cassette. There are no spool cheeks, so the full spool can occupy more than half the space, thus saving space. Tape passes from one spool to the other along the apertures at the side. A built-in pressure pad presses the tape against the head. The drive spindle is inserted through a hole in the casing behind the tape. External pressure roller grips the tape against the spindle.

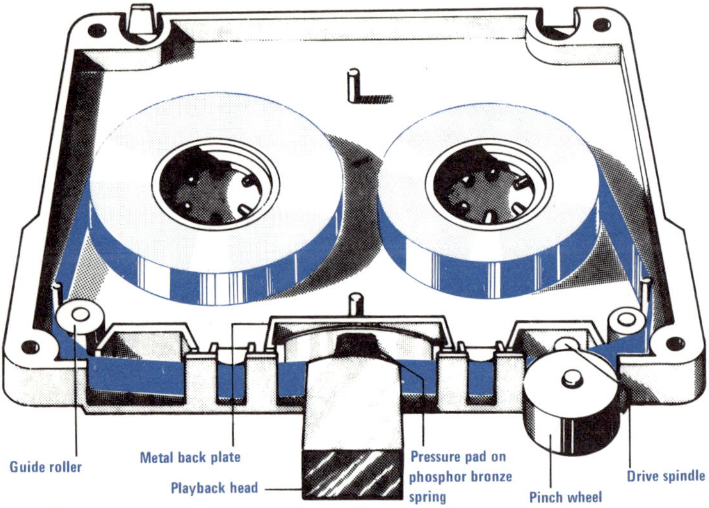

casing, into which the drive spindle is inserted. The rubber pinch wheel is therefore applied to the front of the tape instead of behind it, as with the cartridge, and so it is not included as part of the cassette. As with open-reel machines, the pinch roller is an integral part of the player.

The drive cut-out and hole is duplicated at the other end of the section, because when the tape is turned over to play the other side, the positions must be transposed. A metal backing plate is fixed behind the pressure pad to help screen the head from stray magnetic fields. The tape hubs have centre holes into which the drive capstans of the player are inserted, and they have a number of projecting keys, usually six, which engage with splines on the capstans. As with the open-reel system, the take-up spool is driven by the machine as well as the tape.

Cassette tape

While cartridge tape is 6.25 mm (¼ in) wide, cassette tape is 3.81 mm ($^{1}/_{8}$ in) wide. In spite of being half the size of standard tape, the track widths are three-quarters that of the standard, four-track format, being 0.75 mm (0.029 in) instead of 1.0 mm (0.039 in). This is achieved by reducing the margins between them.

Standard 6.25 mm (¼ in) tape as used on open-reel recorder showing size and disposition of four tracks.

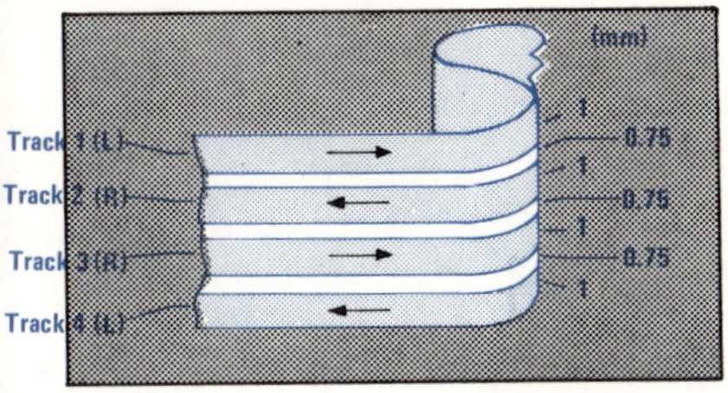

3.18 mm ($^{1}/_{8}$ in) cassette tape, with stereo tracks.

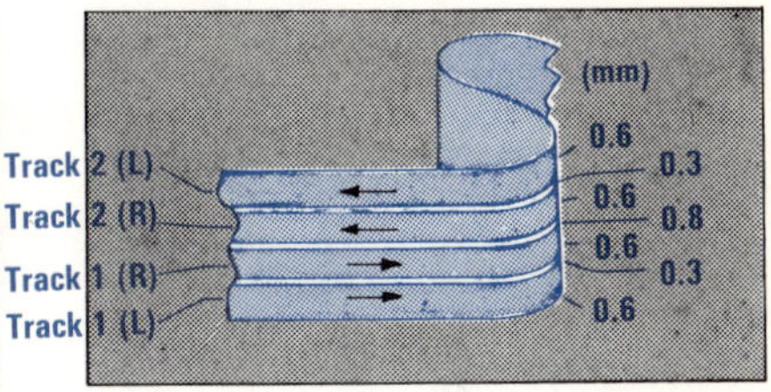

Instead of 0.75 mm (0.029 in) margins, cassette tracks have 0.26 mm (0.01 in) margins. So track width is not much less than standard four-track open-reel systems, and as the cartridge uses eight tracks, a cassette track is substantially wider than a cartridge track.

The significance of this lies in the fact that the wider the track, the larger the signal and the lower the noise level. Also drop-out, which is the momentary cessation of sound due to small irregularities in the tape, is less troublesome.

The tracks are not interleaved as they are with the cartridge, but are adjacent, tracks 1 and 2 being the first stereo pair and tracks 3 and 4, the second. The reason for this is that any cassette can thus be played on either a stereo or mono playback machine. With a mono outfit, the large gap reads both the side-by-side tracks and they are both reproduced as a mono signal, with nothing lost except the directional information. The track pairs run in opposite directions, which of course is a function of the cassette being turned over to play the second programme.

Playing Dolby tapes

When used with high power amplifiers in the home, the noise level of cassette recording can be obtrusive even though the tracks are fully three-quarters the width of those of standard open-reel machines. This is why the *Dolby B system* was introduced. Most prerecorded tapes are now dolbyised and can be identified by the back-to-back double D design on the label. What happens is that during recording all low level, high pitched sounds are boosted, then during playback reduced by the same amount, so that the eventual proportion is the same as the original. The advantage arises in that as noise is mostly low level, high pitched hiss, any such noise occurring in the recording process is also reduced in the playback circuits. Thus the original music sound comes out the same, but the proportion of hiss is reduced.

In the case of car cassette systems wind and road noise masks any tape hiss that may be present, so noise reduction is not of such importance as it would be in the home. Furthermore, Dolby circuits can add greatly to the cost of a cassette machine. They are therefore not usually to be found in a car player. This does not mean that a prerecorded dolbyised tape cannot be used in a non-Dolby machine. They can be, and the only audible difference is that they sound brighter than a normal cassette. If a tone control is fitted, a little treble cut will give a degree of compensation if desired.

The standard tape speed for a cassette is 4.76 cm/s ($1^7/_8$ in/s). This is slow by open-reel standards, and is only half the speed of the cartridge. The principal reason for this speed is tape economy, which means a long playing time on a small amount of tape. This in turn means a small tape pack, which in the case of a car system is an important requirement. However, there are penalties for using such a slow speed. The main one is a restriction of the high frequency response. We saw in Chapter 3 how the higher pitched recorded sounds are related to tape speed and also head gap width. Although present cassette speed would have meant a very poor frequency response not so many years ago, head design has improved and gap widths have been reduced to the extent that the upper response can approach real hi-fi standards. For mobile use when the ear is less critical of these finer points, the upper frequencies are more than adequate.

Another factor dependent on tape speed is wow and flutter, those minor fluctuations which cause pitch variations and distortion. The larger the flywheel mass and the faster its speed, the more effective will be the elimination of speed variations. A slow tape speed, therefore, means also a slow flywheel. It is true that flywheel speed can be, and is increased by using a thin drive spindle, which thus must revolve faster to move the same length of tape in a given time. If carried too far, though, this defeats its purpose as the grip on the tape exercised by a thin spindle is less and therefore slippage and speed fluctuations become more likely. Spacings in the mechanical system preclude the use of a large flywheel such as used with cartridge systems, so in theory wow and flutter should be a lot worse.

In practice it does not quite work out that way. Cassette mechanisms have been developed and improved over the years for the home hi-fi systems, and wow has been reduced to remarkably low figures, only marginally worse in fact than with the cartridge player, even though this runs at twice the speed and has a much bigger flywheel. The more humble car player has thus benefitted from hi-fi technology spin-off. Perhaps if the cartridge had been similarly developed for hi-fi use, it would give the much better performance of which it is theoretically capable.

Tape drive

The mechanical drive system of a cassette player is much more complicated than that of the cartridge, and details vary from make to make and model to model. A belt drive is used from the motor to the flywheel, either driving the flywheel rim or a pulley wheel on the same spindle. Also driven by the belt is a pulley which forms part of the take-up clutch assembly. The slipping clutch arrangement is necessary in order to compensate for the continually reducing speed of the take-up spool as it fills. The clutch assembly drives the take-up capstan via a rubber-tyred wheel.

A further drive is taken from the flywheel through a rubber contact intermediate wheel, pulley wheels and belt to a drive drum, which in the fast-forward position engages the rubber-tyred take-up capstan wheel, and in the rewind position drives the paying out capstan wheel through a small rotation reversing wheel. To simplify matters in some models, especially those having a radio as well, either the rewind or the fast-forward is omitted. The second belt is also often eliminated by having the movable idler driven directly from the flywheel rim. Sometimes too, a separate motor is used for take-up.

Philips cassette drive system with positions slightly distorted to clarify operation. Flywheel and drive spindle are driven via the main belt from the motor. Belt also drives the clutch assembly which drives the right-hand tape capstan. A wheel driven by a small belt from an intermediate pulley, which makes contact with the flywheel rim, is moved to drive either the right-hand capstan for fast-forwarding or the left-hand one through a small rotation-reversal intermediate wheel for rewinding. For normal playing it is as shown, disengaged.

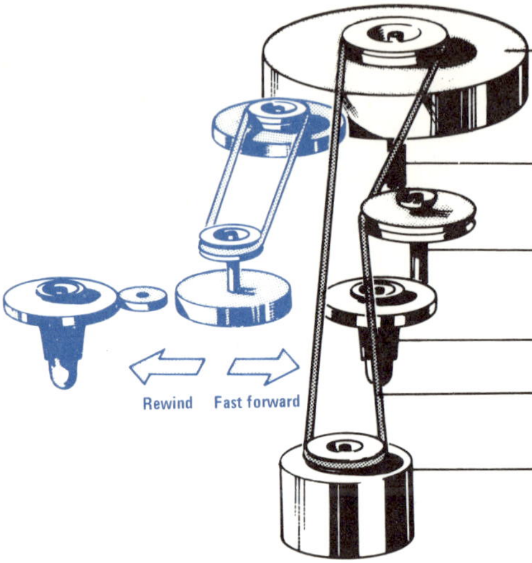

It is not always possible readily to follow the drive sequence when examining or servicing a cassette player, because part of the system is usually mounted above the sub-chassis, and part below, with the linking spindles passing through holes in the deck.

Cassette loading

Now we come to a snag which adds further complication to a car cassette playing system. With the ordinary cassette recorder the cassette is dropped into a well the same size as the cassette, or popped into a swing-out magazine which is then closed into a similar sized well. The reason for this method of loading is that the drive spindle and hub drive capstans must be inserted into the cassette, so the cassette must be placed sideways-on to these components to receive them. The pressure roller and heads are mounted on an arm which moves in to engage with the exposed tape at the edge of the cassette when the start switch is pressed.

This means that the control panel must be at least the size of the loading space, which in turn is the size of the cassette plus room for the controls. Although this is no problem with an ordinary recorder where the cassette is accommodated on the top of the instrument, with a car player it would mean an impracticably large control panel. In the case of the cartridge there is no insertion of drive components and so it can be loaded edgeways into the player.

The only solution to the problem is to use a completely different loading arrangement from the portable cassette recorder and make it so that the cassette can be loaded edgeways like the cartridge, 'mailing' it into a slot in the control panel. This would be impossible with the drive spindle and capstans in place, so the whole drive system, including flywheel and motor, is mounted on a sub-chassis within the player that is pivoted so that it can drop below the level of the main chassis which carries the head, pinch roller and other parts.

When a cassette is inserted in the playing slot, it is secured by a sprung retaining arm, and also actuates a switch which energises a solenoid that lifts the sub-chassis up so that the drive components engage with the cassette holes. When the cassette has been played, the sub-chassis latch is actuated and the assembly drops to allow the cassette to be ejected or withdrawn manually.

An alternative arrangement employed in some models is to slide the cassette between a pair of pivoted guides as it is inserted in the slot. These then drop the cassette on to the drive components permanently mounted on the main chassis. The guides are held in the playing position by sprung roller arms which release them to lift the cassette clear for withdrawal when it is finished.

Cassette ejection

While with the cheaper machines switch-off and withdrawal is performed manually, with most it is initiated automatically at the end of the tape. There are several ways this can be done. It could be actuated by a short length of metallic foil at the end of the tape in a similar manner to the cartridge track change. The snag is that not all cassettes have foil at the end, many just have a blank leader tape, and some do not even have this. Obviously any ejection system must work with all cassettes that are likely to be used.

A common method is to drive a small rocker switch from the take-up hub, so that as long as it is turning the switch generates a steady stream of pulses. These are sensed by a transistor circuit which controls a solenoid. As soon as the pulses cease, when the hub stops, the circuit activates the solenoid and the switch-off and ejection process is initiated. This arrangement has a built-in safety factor. It is not unknown for cassettes to suffer tape tangles which can make a fearful mess of the tape and be very difficult to extricate from the machine. Should the take-up spool become jammed thereby causing a tangle, or should a tangle from some other cause stop the spool as it very likely would, then the switch-off and ejection operation will be triggered before the tangle can become worse.

Automatic reverse

A feature that is to be found with many car cassette players is the automatic reverse facility. When the tape comes to its end, instead of being ejected for the driver to turn it over to play the

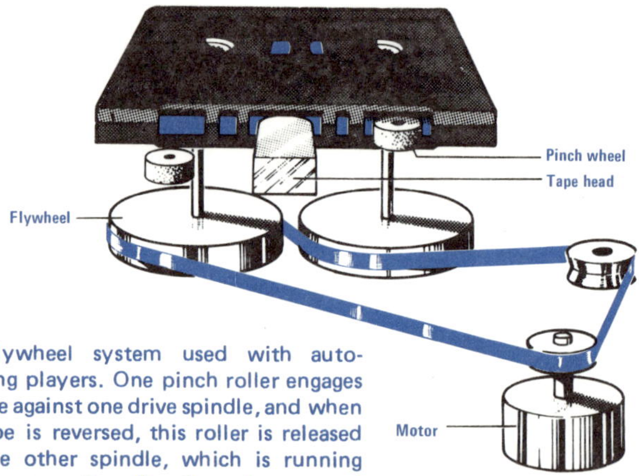

Twin-flywheel system used with auto-reversing players. One pinch roller engages the tape against one drive spindle, and when the tape is reversed, this roller is released and the other spindle, which is running in the opposite direction, is engaged against the tape by the second pinch roller.

other side, the tape reverses direction and the other pair of tracks are played without any attention from the user.

To do this two flywheels and drive spindles must be used which constantly rotate in opposite directions. This is done by driving both with a drive belt from the motor which passes over one half of one flywheel and the opposite half of the other. Both spindles are inserted into their respective holes in the cassette, but only one of the two pinch rollers engage. Drive to the take-up capstan through the clutch is changed from one capstan to the other by means of a wandering idler wheel.

The head has four gaps instead of the normal two, and the appropriate pair are switched in when the mechanical change-over is made. This is accomplished by a similar device to that which triggers the automatic eject. A button is usually included to change manually to the other side before the end of the tape if so desired. A bonus with having two contra-rotating flywheels is that any speed variation which may arise due to making a turn is nullified by cancelling the effect.

Add-on units

As well as complete cassette players and radio/cassette units, there are also cassette units which can be used in conjunction with an existing car radio. They consist of the mechanical equipment, switching and the preamplifier, but without the power amplifier which is provided by the car audio circuits. The same loudspeaker is also used. A connection must be brought out from the car radio audio input circuit, usually the top end of the volume control to the player switching. Insertion of the cassette operates the switch so that the radio is turned off, and the player circuits and motor are switched on.

The dash-board aperture will already be occupied by the radio, so these add-on units are designed to be mounted under the dash. As previously mentioned, most car cassette units are players only, i.e. they will not make recordings. There are some models though that will record as well, and they can be useful for recording memos, dictation and other applications.

Yet another possibility is the mounting bracket with connections to take an existing portable cassette recorder. When plugged into the bracket, the recorder's internal batteries are disconnected and it is powered from the car battery. Playback is via the car radio as with the add-on units, but the additional facility of being able to make recordings as well, and rapidly to detatch the recorder for use elsewhere, makes this an attractive proposition for many users.

6. Cassette or cartridge?

Having discussed the respective features of cassettes and cartridges, now comes the basic question as to which one to go for. There are, by the way, other systems in both the cassette and the cartridge format, but most have either dropped out of the running or were never serious contenders from the start. Some may still be around, but remember that nearly all prerecorded tapes are in the conventional Philips cassette or Lear-Jet cartridge form, so anyone opting for one of the others is likely to have a very limited range of programme.

Performance

Let us then weigh up some of the pros and cons of the two main formats. Potentially, the cartridge should give the better sound reproduction. It runs at twice the speed of the cassette which in theory should mean a better high frequency response and less wow and flutter. However, as we saw in the last chapter, cassette mechanisms and heads have been developed and improved in order to make them contenders with open-reel machines for home hi-fi, so their performance is much better than would perhaps be expected from such an arrangement. In this respect they are rather like the internal combustion engine, really a crude and inefficient device, yet developed over the years to the present state of convenience and reliability.

The cartridge on the other hand has not been developed to the same extent. Although some home cartridge units have appeared, they are few and the cartridge has remained a means of giving background music in the car, a non-hi-fi application. So the performance of the cassette is well up to the standard of the cartridge. Frequency response is about the same and wow and flutter only marginally inferior. Thus a typical wow and flutter figure for a cassette player is 0.35%, while that for a cartridge unit is 0.3% – only 0.05% difference. Frequency response of up to 10 kHz is claimed for both. The actual track width is wider for the cassette than the cartridge even though the tape is narrower. This as we saw earlier is because four instead of eight tracks are accommodated, and the inter-track margins are less. This means that tape noise and drop-out should be less with the cassette, although here again it is doubtful if much difference could actually be heard when driving.

Convenience

We can then forget performance and sound quality when making our choice and go to consider other factors. What about convenience? The cartridge scores here because once it is inserted it needs no further attention, playing all its tracks in turn and then starting again. The cassette has to be turned over at the end to play the other side. This, though, is not too much of a snag as each side can run for up to an hour, while the cartridge will run for up to 80 min before starting again. If we accept the slight task of turning the cassette over, then we can have up to 2 h without repetition, which is more than with the cartridge. It should be noted that these are maximum times – prerecorded cassettes and cartridges are usually less than this.

If a cassette unit with automatic reverse facility is chosen, turn-over is eliminated and the complete cassette will play and start again at the beginning without attention, just like the cartridge, but with potentially longer playing time. So far then, there is little to choose between the rival systems. One aspect of operating convenience which may be worth considering, though, is the respective loading operations. Although both are simply inserted into a slot in the front of the player, the cassette must first be removed from its container. Containers vary in design, but this usually means flapping back a hinged lid and withdrawing the cassette from a recess. Then the container must be closed and put somewhere safe. When changing a cassette, the old container must be opened, the cassette put in the right way round and the case closed, whereupon a new cassette is removed from its container.

All this requires attention and some dexterity if one is also driving, and obviously it constitutes no small degree of risk. The cartridge on the other hand is not kept in a container, since the tape is protected by a slip-on cover. This can easily be removed and refitted without taking too much attention from the road. Ideally of course either cassette or cartridge should be selected and got ready at the *start* of a journey, or the driver should pull over to the side and stop to do it. In this case there is not much to removing a cassette from its container and mailing it in the player; however, traffic conditions may not always make it convenient to stop, and the natural tendency is to drive on one-handed while operating the player. There is really no reason why cassettes should not just have a cover similar to the cartridge instead of the library case in which they are usually housed. Such covers may in fact be made, and if they are, or should become available in the future, they would be a much more convenient way of storing cassettes for mobile use. One disadvantage of the cassette format would thereby be removed.

Reliability

Reliability is an important factor. Once installed, any tape player can be awkward and time-consuming to remove for repairs. The merits of a player that keeps going without trouble are therefore particularly desirable in the case of car systems. The simplicity of the cartridge player makes it much less vulnerable to mechanical failure than the cassette, and it is generally more robust in its construction. For anyone who anticipates putting the player to long hard use the cartridge is therefore likely to be more satisfactory. Not that cassette units have proved troublesome – on the contrary they have a generally good record of reliability; it is just that simple, robust construction must have the edge over more complex mechanisms for reliability.

Cassettes themselves can produce nasty tape tangles, while cartridges seem to be relatively free from this trouble. Cheaper cassettes are more prone to this, where there may be excessive friction. The long play C120 type are also more susceptible, because these tapes are thinner and so more readily loop around the drive spindle should the take-up torque be momentarily reduced. Most prerecorded tapes have shorter playing times and so being of thicker tape are less prone to tangling. Cartridges can deteriorate though through loss of tape lubricant.

Cost and compatibility

Cost is another necessary consideration. Because of their simplicity, cartridge players are cheaper to buy than an equivalent cassette unit. Cartridges themselves though are slightly dearer than the prerecorded cassette carrying the same programme. As regards availability of prerecorded tapes, most commercially issued recordings are made in both formats, so there is little cause for concern that a desired recording may not be available in either one or the other.

One feature which weighs heavily in favour of the cassette is compatibility with other players or recorders. Although as mentioned before, some home cartridge systems can be obtained, most domestic tape installations use the cassette. Likewise all portable recorders are of the cassette type, except for a few that employ open reels. This means that a cassette can be enjoyed at home, on a portable machine or in the car, whereas a cartridge will in most cases be confined to mobile listening.

Making your own recordings

Another related factor is that a cassette system is not confined to prerecorded material as is the cartridge. Some cartridge recorders have been made, but they are rare and expensive. On the other hand, cassette recorders are common and comparatively cheap. We have seen too in recent years the growing popularity of the radio recorder, whereby radio programmes can be taped direct off the air without the fuss of connecting up a separate radio to the recording machine. Recordings thus made are very good in quality, indistinguishable in fact from a prerecorded tape when played on a car system, and almost equal to it when heard over a hi-fi outfit, except for not having Dolby noise reduction. Recordings can be made from a hi-fi tuner and dolbyised cassette deck which are the equal of the commercial prerecorded tape. Just one point to note in the case of the radio recorder – this unit usually records in mono only, although some are stereo too. As discussed in Chapter 2, stereo in the car is much inferior to that in the home, so mono reproduction through the two car speakers will in most cases sound almost as good as stereo.

It is of considerable advantage to make one's own recordings for playing back in the car. For one thing, although the prerecorded repertoire is large, the choice of music and artistes is considerably widened by taking recordings from the radio, or if the law allows, from one's collection of discs. Over a period, quite a good library of tapes can be built up, and of course if one tires of any, they can be re-recorded with something else.

The other advantage is that blank cassettes can be bought quite cheaply, although cheap brands are not recommended. Some of these are prolific producers of distortion, wow and poor frequency response, as well as being mechanically unreliable. Always buy a well known make, which can usually be obtained from discount firms at a good reduction, especially if a quantity such as a dozen are purchased at a time. The cost per unit works out considerably less than that of a prerecorded tape — another very good reason for opting for a cassette system.

Size

Another point to note is the size of the cartridge compared to the cassette, it being roughly twice as big. Storage space in the modern car, especially the smaller models, is not too plentiful, with what there is usually taken up with maps, guides, dusters and other items. Finding space for a collection of tapes is therefore not easy, and cartridges can be particularly difficult to find room for. If taken out of their cases cassettes will occupy even less space, as they are then only about half the width of a cartridge as well as being half the area, so making them a quarter the total volume. However, they should not be stored without some form of protective cover over the exposed tape, and should also have clips in the hub holes to prevent the tape becoming loose.

Summing up then, it can be seen that both formats have advantages and disadvantages, but on balance the cassette's advantages seem to outweigh those of the cartridge. This would no doubt explain the rapidly growing popularity of the cassette. However, this must be a personal choice, and some will feel that the features of the cartridge system make it preferable to the cassette.

Quadraphony

If you are very keen to install quadraphony in your car, then the cartridge is the only answer. However, it should be noted that the list of quad cartridges is quite limited at present, so much if not most of the time you will probably be listening to ordinary stereo cartridges. Furthermore, quad systems are very expensive as there must be four of everything instead of just two. Another point is that quad cartridges have only half the playing time of a stereo cartridge, as four tracks are used at a time out of the eight, in place of the stereo pair.

Cassette-to-cartridge converter

For those who just cannot make up their minds and who want the best of both worlds, there is another possibility. This is the ingenious device which enables cassettes to be played on cartridge machines.

The Belair SCP 580, which enables cassettes to be played in any cartridge player.

An example is the Belair SCP 580 Stereo Cassette/Cartridge Converter. The unit is the same width as a cartridge and with the same shaped front end, but it is longer. It is pushed into the slot in the player just like an ordinary cartridge, leaving a section protruding into which the cassette is dropped. On the front is the single control button for starting and stopping the unit.

The mechanical drive system is similar to that of an ordinary cassette player, with a flywheel/drive spindle and a clutch for the take-up capstan, all inside the unit. There is no internal motor because the drive is obtained from the cartridge player drive spindle. A rubber pressure roller, similar to the pinch wheel in an ordinary cartridge, presses against it and it is coupled to the cassette drive spindle with a gearing ratio to provide the correct cassette speed. There is no rewind, but a fast-forward facility can be obtained by releasing the 'play' control while the cartridge unit is still running. Rewind can of course be obtained by turning over the cassette and fast-forwarding.

Also included in the unit is a preamplifier that amplifies the output of the cassette playback head. This signal is then fed to a transfer head located at the leading edge of the unit to face opposite the playback head in the cartridge player. There is thus magnetic coupling between them, and the signal is transferred by this means into the cartridge player's amplifying circuits. The system is two-channel so that stereo reproduction can be obtained.

Power for supplying the preamplifier is picked up from the cartridge player. It will be remembered that there are a pair of contacts in every player fitted so that as the tape passes over them the track-change solenoid is energised when the end foil appears. A pair of contacts in the converter unit matches with these and so takes off a supply current from them. Although the player contacts are in series with the track-change solenoid, it is not energised by the current taken by the converter preamplifier, because being only a few milliamps it is too small to have any effect.

A voltage stabilising circuit is included to compensate for voltage differences between 9 and 24 V. Thus the unit can be used in any cartridge player without modification.

As might be expected, the performance is not as good as with a good cassette player – frequency response is specified as 40–7 kHz, and wow and flutter a little over 0.5%. However, the audible effects are not great and it is an acceptable compromise if the facilities of both cartridge and cassette are required.

7. Car radios

Car radio requirements

'Why do we need a special radio for the automobile, why not use an ordinary portable fitted up to the dash-board or some other convenient spot so it could be taken out and used elsewhere when needed?' This question is often asked by car owners, and it may indeed seem that in view of the small portable radios now available, a permanently installed car unit would be unnecessary.

There are, however, special requirements and conditions that the car radio must meet, resulting in various design differences. For example, most radios are mounted in the special slot provided for the purpose in the fascia panel and so the station scale and control knobs must all be located on the one narrow side, unlike the majority of radios, where they are found in various positions on the top, sides and front.

The loudspeaker is always internally mounted in the portable radio, usually at the front. If this were done in a car radio it would be buried behind the dash. Instead the speaker is fitted to some place external and often remote from the radio, and if the radio receives stereo broadcasts, or is combined with a stereo tape player, then of course there are two loudspeakers.

If required, the radio can be mounted together with the loudspeaker in a console which is fitted across the gearbox. This method may be the only possible one in cases where the radio cannot be fitted behind the dash or where there is no convenient spot for the speaker. This is sometimes the case in sports models where the fascia is fully occupied with instrumentation, or in commercial vehicles. In the latter case the radio console may be mounted in any suitable position, horizontally, vertically, on the side or front scuttle, or even overhead to the cab roof.

Antennas

All portable radios have a built-in antenna which consists of a high permeability ferrite rod made of small iron particles insulated from each other, on which are wound the input tuning coils. In order that this antenna should not be screened and so reduce the radio waves picked up to a very low level, the radio case must be non-metallic and is nearly always made of plastic.

An internal antenna is not possible in a car, for two reasons. Firstly, the radio is shielded from the radio waves by the car body which of course is metal in the majority of cases. There would be no shielding problem with glass fibre construction, but this type of body is not often found. An external antenna fitted outside the car body is therefore essential, or at least one mounted above window level where the shielding effect is less. Operating a ferrite antenna inside the car results in very weak reception with a lot of noise and fading, as those who have tried to use an ordinary portable will have discovered. The fading is due to the directional characteristics of the ferrite rod. Its pick-up is at a maximum when broadside-on to the transmitter. As the car changes direction continually during the course of a normal journey, signal strength varies so causing fading.

Many portable radios have a car-antenna socket so that they can be used in the car, and an external antenna mounted on the car body just plugged in. With some models the socket feeds a coil on the ferrite rod, thereby injecting the picked-up signal, but with the better ones the ferrite antenna is completely switched out and the socket is switched in to the input circuits. While the latter arrangement is the best, neither is really satisfactory, which brings us to the second reason why the portable radio is not suitable for use in a car.

Considerable interference is generated by the ignition system, and to a lesser extent by the various electrical accessories such as windshield wipers and clock. As the interference fields are greatest in and around the dash-board wiring it can be seen that a radio fitted into the dash is in a very vulnerable position. So, a ferrite antenna here will certainly pick up more noise than radio programme. Not only the ferrite rod but also the receiver circuits themselves pick up interference. This is why the portable with the car-socket which switches out the internal ferrite antenna will pick up interference as well as the one where it remains in-circuit, although to a lesser extent.

Shielding

The only way to cut out interference (apart from fitting suppressors to the interference sources, which must be done anyway) is to house the radio in a metal box and thus shield its circuits from stray interference fields. So car radios usually have steel cases which must be securely fitted and all the fitting screws screwed tight home to ensure that all parts of the case are grounded. The case itself must then be well grounded to the car body; in most cases the fixing bolts will do this, but if mounted in a part that is insulated, e.g. a wooden fascia, a separate grounding braid must be run to the nearest metal bodywork.

As the wire from the external antenna to the radio is also vulnerable to pick-up interference, it is shielded, but this raises another problem. Shielded cables have the property of *capacitance*, i.e. they behave as though they were a capacitor, and this capacitance increases in proportion to the length of the cable. Now any capacitance across the input circuit of a radio will detune it and put it out of alignment with the other receiver tuned circuits. The makers allow for this to some extent, but they have no way of knowing just how long the antenna cable is going to be. In some instances it may be short, where the antenna is mounted on a front wing, but in other cases it may be long, such as when fitted to the rear of the body. Furthermore, different cables have different capacitance values. To overcome this the radio is fitted with a trimmer which consists of a variable capacitor. This compensates for the varying antenna cable capacitances, and must be adjusted with the car antenna plugged in, to give maximum volume. Adjustment is accomplished by a screwdriver through a hole in the casing.

Mounting

Modern radios are compact and there is little difficulty in mounting them. Not so the older models though, especially those using vacuum tubes, or the hybrid ones employing tubes and transistors. These were much larger and it was necessary to support the rear of the unit as well as the front by one or more supporting struts to the bulkhead. Some models were made in two sections, the radio with the control panel and the power unit comprising output transistor and driver stages. Older all-tube units often had the vibrator, transformer and rectifier mounted in a separate unit.

Connection between the units was made by a multicore lead with a four- or five-pin plug on the end and a corresponding socket on the power unit. Battery supply leads and loudspeaker leads were also taken from this unit. Where there was room behind the dashboard the power unit could be bolted to the back of the radio to make one piece, but otherwise a mounting position elsewhere had to be found for it. Fortunately, transistors have eliminated the need for such bulky units, but they may still be around in some older models, and no doubt giving good service.

Supply voltage

Another difference between the car and the portable radio is the supply voltage. Portables run on 6 or 9 V as a rule, but the car battery is rated at a nominal 12 V. As the charging voltage from the generator must always be higher than the battery and this is applied on-and-off as the car motor is running, most car radios are now rated at 14 V.

Power output

Portable radios also have a very limited power output because of the necessity of operating them from dry batteries which have a restricted output current. A compromise has to be reached between adequate power and reasonable battery life. Powers of up to 1 W are usual, generally more than enough for the application. With car radios there is a high ambient noise level if the vehicle is travelling at speed. Motor noise is usually a lesser component in this, and wind and tyre noise together with that of other traffic make up the major portion. The total noise is not very obtrusive as it is steady, and driver and passengers soon get used to it; as a result they may not be aware that it is as loud as it really is. When
 using the radio it is therefore found that a much higher volume

level must be set than if the car was at rest, if broadcast speech is to be intelligible.

It is necessary then for a car radio to have a higher power output than a radio used in much quieter surroundings. This is another reason why a portable used in a car is not as a rule very satisfactory – the volume must be turned well up toward the distortion level. Fortunately there is no supply current problem as the car battery is capable of delivering large currents and is constantly being recharged. Car radios therefore usually start at around 3 W and go on up to over 10 W.

Output Transistors

To achieve these powers, large output transistors are used of the chassis-mounting variety instead of the small wire-ended type usually employed in the portable radio. These generate a considerable amount of heat and must be so mounted that the heat is dissipated. A common way of doing this is to fit them on the outside of the rear radio-chassis member with the connecting pins passing through holes for connection to the circuit inside. The chassis member is of thicker gauge metal than the rest, and is

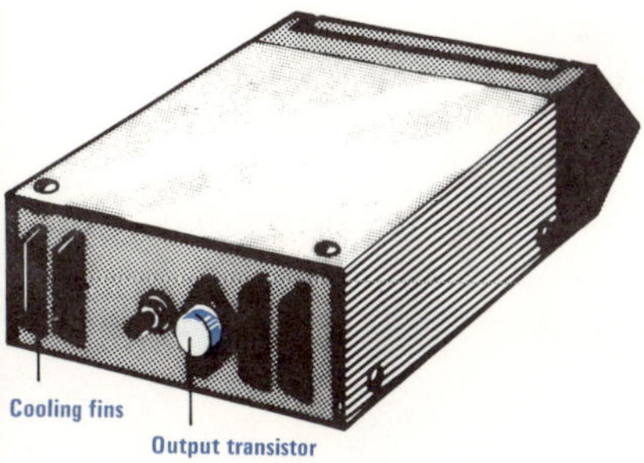

The output transistor is usually mounted on the back section of a car radio so that heat can be dissipated. Cooling fins are often moulded on to the section.

provided with cooling fins in the case of the higher powered radios. The metal transistor casing also serves as one of its electrodes and so it must be insulated from the chassis. This is done by interposing a mica washer between the transistor and the chassis, which readily conducts heat but insulates electrically. If the transistor is exposed, care must be taken when installing the radio that no metal part of the car body shorts against it, thereby grounding it. Most radios have a plastics cover which protects the transistor against any such possible shorts.

Transistor output circuits can be either of the Class A or Class B type. Class A uses one or two transistors drawing a large current from the supply to keep the operation on the linear part of the transistor characteristic curve. Class B employs two transistors running at low current and therefore operating on the non-linear portion of the characteristic. However, the circuit is so arranged that the distortion thereby generated by one transistor is cancelled by that of the other. We thus get a high output with low supply current, but the snag is that some distortion remains. So in spite of the high supply current, and incidentally also the greater amount of heat to be dissipated, Class A is often chosen by the designer for its low distortion. All portable radios have Class B output stages in the interests of battery economy, but car radios vary, some using Class B while others operate with Class A. The standing current with no signal, i.e. the volume turned right down, is about 50 mA in the case of Class B, but between 500 and 750 mA with Class A, so it can be seen that the current difference is appreciable.

A.M. or F.M.

Radio waves can be either amplitude modulated or frequency modulated, that is the carrier wave can either have its height varied in sympathy with the sound or its frequency. Both a.m. and f.m. transmissions are broadcast, and although a.m. is the most common, many portable and domestic radios are designed to receive both. Car radios likewise are available that will pick up both types of transmission, although the majority will receive a.m. only. There are several difficulties involved in receiving f.m. in the car, which is why the facility is less common than with portable radios.

One of the major snags is that of interference. While f.m. is free from many types of electrical interference that affect a.m. sets, ignition interference does have frequency modulated components and so affects f.m. reception. The effect is worse than that produced on a.m. radios and interference suppression is more difficult. Now while it is possible to suppress interference by the car in which the radio is installed, the antenna will still pick up interference from every passing vehicle that is not f.m. suppressed. So there can be considerable crackle of interference from other vehicles.

Multipath Distortion

Another problem is the peculiar form of distortion that affects f.m. reception known as *multipath distortion*. This is caused by reflected signals arriving at the antenna slightly later than the direct one, owing to the longer path travelled. On TV receivers the effect is known as *ghosting* and can be seen as a succession of subsidiary images displaced to the right of the main one. Owing to the much longer wavelength of a.m. signals, such delays have negligible effect, but the high frequency f.m. signal is very prone to it.

Ordinarily, multipath distortion can be reduced or eliminated by altering the antenna location or orientation to cut out the pick-up of secondary reflected signals. Obviously this cannot be done with a moving vehicle, and although the multipath reception would soon pass as the vehicle's position changed relative to the sources of reflection, there would be others continually coming into range as the journey proceeded.

The range of the f.m. station is much less than that of the a.m. transmitter, especially on medium wave, so on a long, high speed journey the car will be passing from good to fringe reception areas quite rapidly when receiving f.m. This will mean frequent re-tuning. Re-tuning will also be required in the case of a.m. broadcasts, but as their service area is much greater, especially the high power network stations, it will be needed less frequently.

It can be seen that f.m. is not really well suited for mobile reception, and so should not be taken too seriously. If an a.m./f.m. car radio costs much more than a comparable a.m.-only model it is questionable whether the extra money would be well spent. One factor though which may swing the choice in its favour is that stereo broadcasts are only transmitted on f.m., so that if stereo radio is desired, f.m. reception will be essential. Not of course that all f.m. radios can receive stereo – they must be equipped with a stereo decoder, two audio channels and loudspeakers!

Station tuning

Having mentioned the tuning of f.m. signals, we must now turn to the subject of station tuning in general. All sorts of station scales are to be found on ordinary radios, the circular or semi-circular type being very common. These are not very practicable for car radio considering the size and shape of the control panel, so the horizontal linear scale is universally used.

With the cheaper radios, or those combined with a tape player, we have a simple scale and tuning knob, but more expensive models include a number of pushbuttons that can be pre-set to any desired station. Combined radio/tape players do not usually have them because of lack of space on the panel.

Individual station selection can be accomplished by switching in one of a number of tuned circuits, and in fact this was the method used in many of the old vacuum tube radios popular some years ago. Actually at least two tuned circuits were needed for each pre-selected station in addition to those required for the normal manual tuning, and each needed an adjustable component in order to tune in the desired station. All this was very complicated and took up a lot of room, and so not very practicable for a car radio. Instead a mechanical system is usually used.

Mechanical Pushbuttons

With this, each button actuates a long threaded rod which carries a striking plate. When the button is depressed, the plate is pushed against a spring-loaded thrust bar linked to the tuning device and holds it in the selected position. A latching mechanism keeps the rod in place. When another button is pressed, the latch is released, allowing the previous button to return under the power of its return spring, and the striking plate on the new button shifts the thrust bar and also the tuner to a different position.

The button is adjusted by firstly pulling it outward, thereby engaging it with splines on the rod, then turning it, so causing the striking plate to be screwed up or down the length of the rod threads. Whenever the button is subsequently depressed the tuner will be moved to the same position. Normally the button is disengaged from the splines so that accidental rotation will not alter the selected tuning position.

The resetting accuracy of these mechanical systems varies according to the precision of manufacture and the amount of wear. Poor accuracy usually shows up by the tuner stopping in a different place when coming from one end of the scale than from the other. So, whether one hits the station spot-on depends on which button was depressed beforehand. Normally though, accuracy is good, and at worst the station will be received slightly off-tune, needing only a slight adjustment from the manual control to bring it in.

Basic pushbutton mechanism. When button is depressed, its striking plate pushes the thrust plate to a predetermined position against the return springs, and a latch engages with a ledge in the button spindle to hold it in place. The thrust plate is linked to the tuner coil cores. Pressing another button lifts the latch releasing the first button and its striking plate pushes the thrust plate to a new position. Striking plates are threaded on to the pushbutton spindle and are adjusted by turning the pushbuttons.

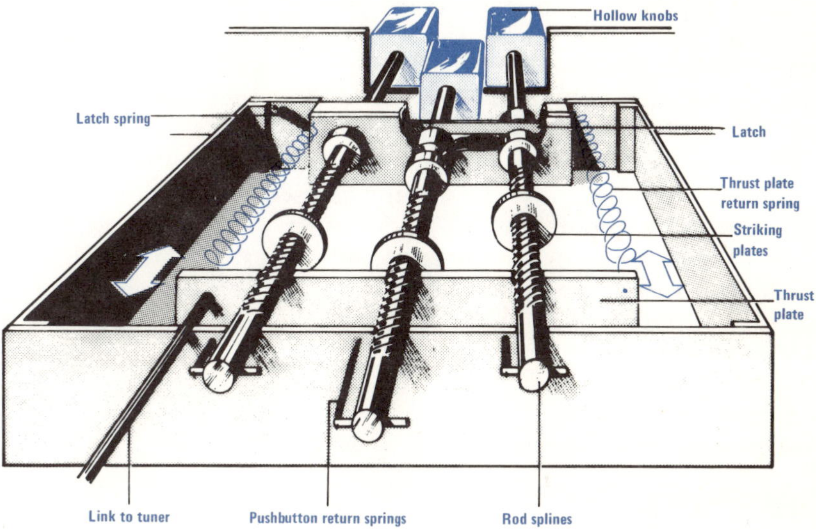

Another method of setting the pushbuttons is to tune the station in by means of the manual control, and then press the button. This sets a selector on the button rod so that further depressions will return the tuner to the same place. The snag here is that any alteration of the tuning knob while the button is depressed and the desired station is being received will alter the setting. Some models overcome this by taking the tuning drive from the manual control through a clutch. Pressing any button opens the clutch, thus disengaging the control knob. With these sets the station must be accurately tuned manually before depressing the button to set it, because once the button is pressed the manual control cannot be used to give the final touch to the tuning. There are other methods of button tuning, but the basic idea is very similar to those described here. In theory any number of buttons could be provided but usually five are given.

Novel station selection systems

Another tuning system is the *station seeker*, whereby the pointer is driven along the scale by a small motor. When a station of usable strength is received, the motor stops; if the station is not required, a button is pressed and the motor drives the pointer on to another station. On reaching the end of the scale, the motor automatically reverses and drives the pointer back over the scale.

In a further system, similar to the pushbutton method, a number of different stations can be mechanically set, but only one button is used to do it. The stations are arranged in sequence so that pressing the button selects the next station. Thus if number 1 is tuned in but number 6 is desired, the button must be pressed five times to go through the intermediate stations. After 6 the next one up is 1 again and so the sequence repeats. There is no means of identifying the stations other than their spacing from any known one, and although only one button is used, which can therefore be operated without looking, attention is more likely to be distracted than with conventional pushbuttons. Certainly though, pushbutton tuning is a big advantage in a car radio, as it can be decidedly hazardous trying to search for a desired station manually while driving.

Permeability tuner. A carriage plate moves the cores in or out of the coils. It travels along a worm drive which is driven by the tuning knob. Individual cores are aligned by adjusting screws.

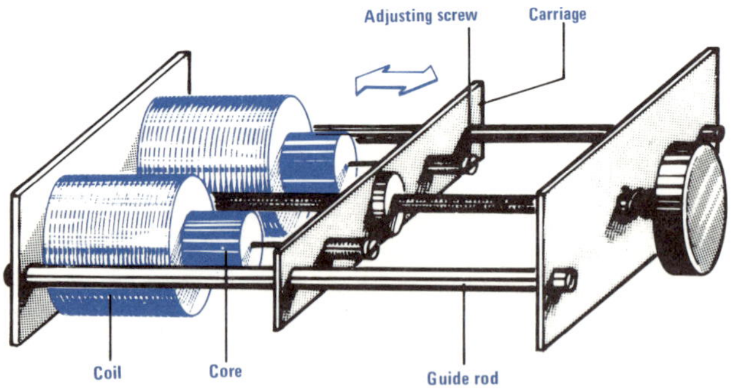

The actual tuning device usually used is a set of two or three coils with sliding ferrite cores suspended by stiff wires from a carriage running along the threads of the tuning rod. This is known as a *permeability tuner* and it facilitates the simultaneous tuning of several tuned circuits. To achieve extra sensitivity car radios often use an r.f. amplifying stage, thus necessitating at least three tunable circuits. Most portable radios, except the more expensive ones, do not have an r.f. stage and so require only two tuned circuits. These are made tunable by means of a two-gang condenser. Here then we have another basic difference between the car radio and the portable.

Input protection

There is a hazard which can afflict the car radio which is less likely with the portable. When the vehicle passes near or under high voltage power lines, a high electrostatic charge can be picked up by the antenna, and with certain circuit configurations this can discharge through the first transistor and thereby destroy it.

Various means have been employed to protect the transistor from this possibility. A resistor shunted across the circuit helps by providing a path to discharge the charge, but if it is too low in value it will also reduce the signal, while if too high it will give insufficient protection. A reverse-biased diode is another means of protection. This has a small voltage applied of opposite polarity to its normal operation, thus making the diode non-conductive and so having no deleterious effect on the signal. Any high voltage in the circuit overcomes the bias, causing the diode to conduct and thereby passing the voltage harmlessly to ground. A snag here though is that the diode itself may be destroyed in the process, thus leaving the circuit unprotected against the next occurrence.

A small neon lamp has also been used as a protection device. These lamps are virtually open circuits at low voltages, but when a certain voltage level is reached they strike over and a current is passed. They have, therefore, no effect on the low voltage signal but strike and conduct when a high voltage is induced. Another method is what is known as a *voltage dependent resistor* (v.d.r.). This is a resistor which alters its resistance according to the applied voltage. At low voltage the resistance is high, but rapidly decreases as the voltage rises. While these technical details may not be of much interest to the user, he should make sure that any radio he buys has in fact got some form of protection against antenna induced voltages.

Car-portable radios

We started off this chapter by showing that ordinary portable radios are unsuitable for use in the car. Some makers have produced models which attempt to get the best of both worlds, and have produced what are known as *car portables*. Basically they are portable radios with ferrite antenna and internal dry batteries. The shape and control panel is that of the car radio, but they are housed in a plastics case.

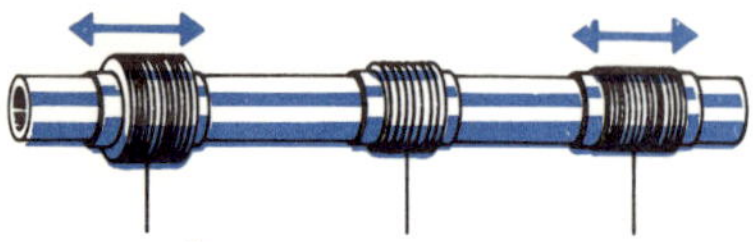

A ferrite rod antenna as used in most portable transistor radios. End coils are for a particular waveband, the middle one is a coupling coil which is connected to antenna socket for use with a car antenna. These are vulnerable to ignition interference though, and portable radios are not recommended for car use.

At the base of the radio is a row of sockets, and the whole thing slides into a metal tray which is installed in the car. The sockets then mate with pins at the bottom of the tray, and a metal finger enters a hole in the radio and operates a switch. This cuts out the ferrite antenna, switching in an external one, and also disconnects the internal loudspeaker in favour of a larger one mounted in the car. A supply from the car battery is switched on in place of the internal dry cells. Metal foil inside the plastics case and the metal tray serve to shield the circuits from interference.

So, the radio can be used either as a portable or in the car, although the output power is restricted when used in the latter mode. A locking arrangement is usually provided to secure the instrument in the tray and thus prevent theft. In spite of the convenience of the system, car portables tend to fall between two stools and are generally less satisfactory than either a separate portable or a car radio.

Vacuum tube car radios

With modern transistor car radios there is little complication involved with the power supply. Transistors operate at low voltages and the 12–14 V car supply is ideal. It was not always as straightforward as this when tube-type radios were in vogue. Vacuum tubes need 100–200 V supply and obtaining this from a car battery needed some ingenuity. While a transformer can be used to step up a voltage to any desired level, it will only work with a.c., so the d.c. of the battery meant that something more was needed.

Firstly, then, the d.c. supply had to be converted to a.c., by means of a vibrator. This worked like an electric bell; a small coil generated a magnetic field when the current was applied and attracted a sprung arm carrying sets of contacts. When this happened the contacts carrying the supply to the coil were opened and the coil was de-energised. The arm returned to its rest position only to close the contacts and energise the coil once more. Thus the arm, like the striker of a bell, was set continually vibrating to and fro.

Another pair of contacts switched the battery supply from one end of a transformer winding to the other as the arm vibrated, thus providing it with alternating current. The transformer then stepped it up to the required voltage, but now it had to be converted back to d.c. This was done either by a rectifier tube or a further set of contacts on the vibrator arm, which were obviously synchronised with the first pair.

The snags with vibrators were that they generated interference and had to be carefully screened from the rest of the radio circuits, usually by enclosing them and the associated components in a steel box, and all supply and output leads had to be filtered with coils and condensers; also the contacts got burnt in time and the vibrator had to be replaced. The vibrator circuit along with the tube filaments were rather greedy for battery current, drawing several amps from it. Indeed, the transistor has solved many problems but nowhere more than in the car radio!

Some radios were made, and may still be in use, that were hybrids, using both vacuum tubes and transistors. The output circuit consisted of one power transistor operating in Class A or two in Class B, but the receiver circuits used a specially developed range of tubes designed to operate from a 12 V supply. The reason for this arrangement was that the transistors then made were not as sensitive and stable at radio frequencies as tubes. The result was a reliable compact unit that did not require a vibrator, although not as compact as modern all-transistor radios.

Interference and polarity

Although the supply to the transistor car radio is straightforward, being taken directly from the battery, some attention has to be paid to the possibility of pulses of interference from other car equipment reaching the receiver through the battery wiring. To prevent this, interference filters are included in the radio supply line. These comprise series chokes and shunted capacitors, the chokes offering a high impedance to transient current changes while not affecting the steady d.c., and the capacitors bypassing any surges to ground.

One matter of considerable importance to the user, both in regard to radios and tape players, is the polarity of the unit and that of his car. One terminal of the car battery is grounded to the bodywork so that only one wire need be run out to the various circuits, the return path being made by the metal work. Radios are therefore designed in a similar manner, with one pole of the supply circuit being grounded to the casing of the unit. It is essential that the same pole is grounded as with the car battery, otherwise the transistors will be destroyed. Over the years there have been changes in the practice of grounding car batteries, and also the standard grounded pole has varied in different countries, so that imported foreign models may be different from home-produced ones.

To take care of this, many radios are made with dual polarity, and the change from positive to negative ground or vice-versa can be made by changing over an internal switch or re-soldering links across a terminal strip. Not all are made this way though, and tape players in particular tend to be made in just one polarity. So before purchasing or attempting to install any radio or tape player in the car, make sure that the polarity is right. More will be said on this subject in Chapter 10.

Commercial vehicles

In petrol tankers there is a serious danger of fire or explosion if one of the battery poles is grounded, as any accidental short from any part of the vehicle wiring to the body could cause a spark. So the battery supply is run with a two-wire system isolated from the bodywork. Should an ordinary car radio be fitted it would ground one of the battery poles and so cancel the protection of the two-wire system. Some makers manufacture a special model for tankers which has an isolated case, and one of these should be used for any radio installation in one of these vehicles. Paraffin (kerosene) is not so dangerous and tankers transporting it can have one battery pole grounded. An ordinary car radio can therefore be used.

Larger commercial vehicles usually have higher voltage batteries than the domestic car. Car radios can be used, though, if some means of reducing the voltage is provided. A simple resistor can in theory be wired in series, but the voltage drop thereby produced will vary in proportion to the current drawn by the radio.

The value of the resistor will therefore depend on the type of radio and its current consumption. With Class A output stages, the current is steady and there is no problem, but with Class B the current constantly varies in sympathy with the signal, so the voltage also fluctuates. It can be steadied to some extent by connecting a large value electrolytic capacitor of around 10 000 μF from the junction of the resistor and the radio power lead to ground.

Anyone conversant with electrical theory and Ohm's law in particular can work out the resistor value for any particular radio taking a current value of about 1.5 times the no-signal value for Class B radios. The wattage will have to be adequate, a wire-wound 20 W resistor being suitable in most cases, and the resistor which will get hot will have to be mounted clear of anything which could be damaged by heat. Readers not so conversant would be better advised to seek the services of an electrician.

An alternative is to use a stabilised supply unit which employs transistors to regulate the voltage over a wide range of currents. Technically this is the best method but also the most expensive.

Going in the other direction, it may be necessary to run a 12 V radio from a 6 V supply. There are still some models of car that use a 6 V battery as well as motorcycles (there is no reason why a car radio should not be fitted in a side-car even if it would pose mounting problems on the bike). For these, the only answer is a voltage converter. This is a unit that works by converting the d.c. to a.c. by means of a transistor oscillator, feeding the result into a step-up transformer and then rectifying it by a pair of silicon diodes. The principle is the same as for a vibrator, but transistors are used instead. These are quite efficient and do not generate interference as did the vibrator.

8. Car antennas

Requirements of an antenna

The function of a receiving antenna of whatever type is to convert the electromagnetic radio waves travelling through space into a corresponding electrical signal that can be applied to the input circuits of the radio receiver. Although it is a simple piece of equipment basically, there are several necessary requirements, as it serves a very important function.

Firstly it is essential that the electrical signal produced is as large as possible so that random noise signals generated in the receiver circuits will be very small in comparison. Where this is not so we get a continual 'mush' in the background plus occasional bursts of static. The antenna must also be effective over the full range of frequencies it is intended to cover, otherwise one band will be lively with plenty of stations coming through strongly, while another, or even the opposite end of the same band, will be comparatively dead. It should be mentioned, though, that in most cases where this happens the trouble will be in the alignment of the receiver circuits.

A further desirable feature is that the antenna should discriminate against noise and interference coming from outside the receiver, although this has more to do with the positioning of the antenna than with the antenna itself.

The simplest form of antenna is a straight piece of wire, but there are a number of factors that govern its effectiveness. One is length, the theoretical optimum being one quarter of the wavelength of the desired signal. If this length is exceeded, the signal picked up by the extra wire is out of phase with that in the rest of the antenna and actually reduces rather than adds to it.

The dipole

A means of getting more signal is to use two quarter-wavelength sections end-to-end, the feeds to the receiver being taken from the two inner ends. These are applied to opposite ends of the input coil in the receiver, so that although they are of opposite phase, the signals are added. This type of antenna is known as a *dipole*, and is the basis of most TV and domestic f.m. antennas. As it is tuned by its length to one particular frequency, it is most effective at that frequency, though nearby stations on the band are received almost as well because the antenna tuning is broad.

If we now take a look at some of the dimensions we can see why dipoles are not normally used for the broadcast bands. For example at 30 m a quarter-wavelength would be nearly 8 m (25 ft), so a dipole would stretch to 15 m (50 ft). At 300 m on the medium waveband, a quarter-wavelength is 75 m (246 ft) and a dipole would be double that! Obviously even single tuned sections are impracticable, and the length of any car antenna is only a fraction of the actual wavelength. In the case of f.m. broadcasts on the v.h.f. band, antennas may approximate to the quarter-wavelength. At 90 MHz for example, the quarter-wavelength is 832 mm (2.73 ft), which is near the length of an extended car antenna. So, for the normal a.m. bands, the longer the better within reason, but for v.h.f. reception the optimum length corresponding to the quarter-wavelength of the received signal should not be exceeded.

Polarisation

Another factor is that in theory the plane of the receiving antenna should be the same as that of the transmitter. This is termed *polarisation*, and it is more important on the very short v.h.f. wavelengths. In practice, the radio propagation conditions tend to twist the plane of polarisation, and to some extent the direction of the car antenna is governed by practical considerations such as mounting and appearance. This then is not a matter that need concern the car owner too much. If v.h.f. is to be received and it poses no major mounting problems, the antenna can be positioned in the same plane as the local transmitters. With the normal broadcast bands, polarisation makes no noticeable difference to reception. For radios that will tune to both a.m. and v.h.f. bands mount the antenna for the best v.h.f. reception, and in most cases the a.m. will come through loud and clear too.

Shielding

Results are much affected by the shielding effect of large areas of metal. This applies to the car body itself, so that antennas mounted inside the car are rarely satisfactory. This is why, as we saw in an earlier chapter, portable radios with built-in antennas give poor results in the car. There is only one type that will perform reasonably well inside, and that is the one that consists of a thin strip of self-adhesive conductor fixed along the top of the windshield. Since glass is transparant to radio waves it does not shield the antenna from the waves, although the roof and door pillars do have some effect. The main problem is that of interference from the electrical wiring to the instrument panel, and for this reason an exterior antenna is much the best.

Shielding can also be caused by large external objects. It is a common experience when going through a tunnel, or even under a bridge, for the radio to blank out completely. In the centre of most towns reception is noisy and difficult due to being surrounded by large metal-framed buildings. In mountainous country, too, fading can be experienced. If listening to your favourite programme when approaching one of these situations, be prepared to miss part of it or make a detour!

The shielding effect of the car body can be put to good use in reducing interference from the motor or from those of other vehicles. This is done by the judicious siting of the antenna. We will cover this aspect in greater detail in Chapter 10.

Antenna types

What types of car antenna are available then, and what are their features? Under-body units were at one time quite popular, but are rarely if ever seen today. They consisted of a rod or a number of rods mounted under the auto. Although a good length was thus attained, they were shielded by the body and were obviously in the lowest position, whereas an antenna needs height. Really they operated more as a capacitance antenna because of the capacitance between them and the ground. Another type which seems to have gone out of favour is the ring, usually of between 150 and 230 mm (6 and 9 in) in diameter, mounted on the roof just over the windshield.

Simple antennas needing little in the way of installation obviously have their attractions, but usually have drawbacks. We have the windshield type mentioned earlier, and there is also a model that clips on the side window. When the window is raised the clip is trapped in the upper window recess and the antenna, which is a short vertical rod, is carried on the outside. The connections pass through the window clip in the form of flat conductors and are terminated in a short length of shielded cable fed around to the back of the radio. The snag with this type of antenna is the trailing wire from the window, which at best is a nuisance and at worst could be a hazard. While the fixing is secure when the window is closed, it can easily be dislodged and fall off when the window is opened. Really then, there is little to recommend it, except perhaps as a temporary measure in a car that is soon to be traded in.

Whip Type

An antenna which has been growing in popularity is the whip type. It consists of a long flexible but springy section which is mounted permanently on some part of the bodywork. The visual effect is to give a rather 'sporty' look, which is perhaps why they are in favour with younger car owners. The length is somewhat excessive, which can create problems when driving in and out of the garage, although from the reception viewpoint this is an advantage. Instead of letting the antenna stand straight up as a cloud tickler, some find it more convenient to bend it back and secure it to the roof guttering with the special clips available for the purpose. As most models have plastics insulation over the whole length of the whip, this does not short the antenna to the car body. Being springy and bendable, it is not easy deliberately to damage this type of antenna, a consideration in these days of increasing vandalism.

Most car antennas take the form of a telescopic set of rods which can be collapsed down to the length of the lowest one or extended to maximum length as required. Within this basic type there are a multitude of different models, but the main differences are in mounting, retraction and angular details, as well as finish.

Vertical Mounting Semi-retractable Type

The vertical mounting semi-retractable (A) is perhaps the most popular type. It is designed for mounting on a horizontal surface, but most are made so that the angle can be adjusted to accommodate a sloping surface, or to give the antenna a rake if so desired. Some models permit a greater angle of deviation from the vertical than others.

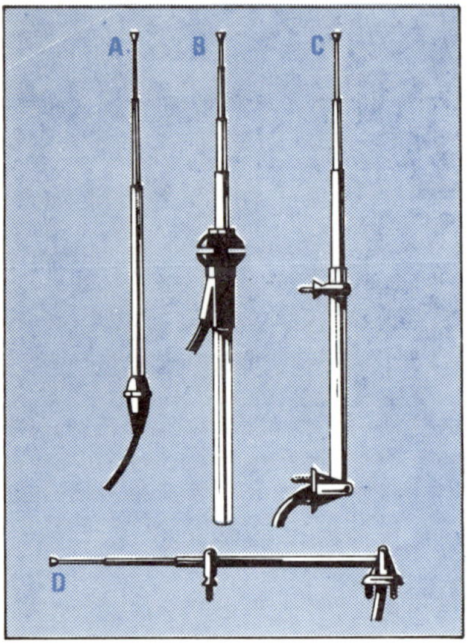

Various types of car antenna. A Semi-retractable wing mounting. Can be mounted almost anywhere, as little underside clearance is required. Retracts to length of lowest section, and can be inclined up to 90° maximum angle. B Fully retractable wing mounting. Less temptation to vandals, especially the lockable ones. Good underneath clearance is needed. C Side/pillar mounting. These have an extra insulated support at the top of the first section to prevent it being displaced and fouling the door. D Roof top antenna, similar to the side mounting type but smaller.

These antennas are mounted through a hole cut in the car bodywork, and as no contact can be permitted between the antenna and the metal, a solid insulator passes through the hole to support it. The feeder is connected inside the insulator, or to terminals at its base. These connections should be fully waterproofed to prevent trouble from moisture thrown up from the wheels; any model without such protection should be firmly rejected. A dome made of metal or plastic covers the top part of the insulator. A ball-and-socket arrangement is often used within the dome to give the variable angle facility.

Projection below the mounting surface is just for a couple of
 inches or so, and retraction is to the length of the lowest section,

but when extended the length can vary, to up to 2 m (6 ft) with some models. For normal operation, a couple of feet extension is sufficient, but if receiving in difficult terrain or trying to pull in a weak station, the antenna can then be pulled out to its full length; the extra length can make a considerable difference under these conditions.

Fully Retractable Type

Unfortunately, the car antenna is an open invitation to vandals, and once bent, the telescopic sections are ruined as little can be done to restore them. For this reason the fully retractable antenna

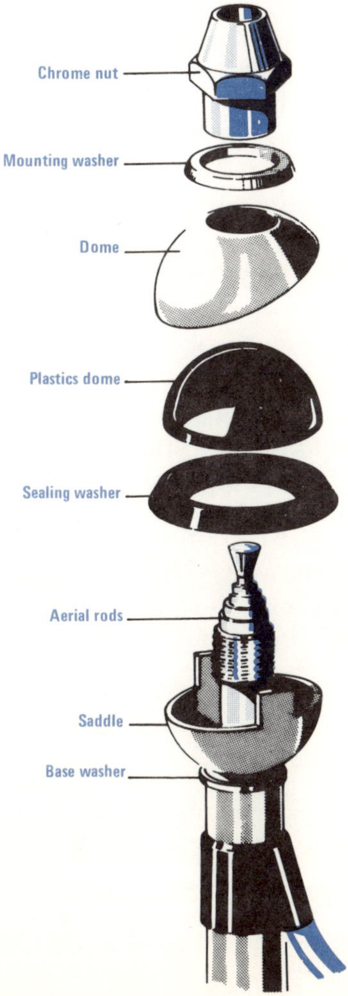

Typical mounting arrangement of a fully retractable antenna. It is essential that good contact be made between the antenna housing and the car bodywork, so paintwork must be stripped off to the bright metal immediately around the hole.

(B) (page 78) has grown in popularity. When retracted the whole mast disappears below the surface, leaving only a small knob to pull it up with. Without the visual stimulation of a slender, vulnerable rod, many potential vandals ignore it. There are some though who will pull up even this antenna in order to wreck it, no doubt considering the extra effort a challenge, so a further protection is afforded by the lockable antenna, a fully retractable model that can be locked in the retracted position.

An important factor when considering fitting a fully retractable model is the clearance available below the mounting surface. Space required varies with different models but anything from 230 to 380 mm (9 to 15 in) will be needed. Certainly, no holes should be drilled before this point is checked. The angle of inclination cannot be as great as with the semi-retractable unit. The latter can be up to 90° from the mounting surface, but fully retractables can usually be varied only up to around 40°. Some models are available that will give a greater fixed angle. Cable entry for fully retractable antennas is not at the bottom of the unit, but at the side just below the mounting insulator. The cable does not therefore form a vulnerable loop, but can be run along the underside of the mounting surface.

Vertical Type

If it is desired to mount the antenna on a near-vertical surface, although the 90° semi-retractable could be used, there are special models made for the purpose. It is possible that an ordinary antenna could become displaced and thereby interfere with the opening of the door. The vertical models therefore have a further insulated support at the top of the lowest section to hold it securely in place (C, page 78). They can thus be used on the sides or window pillars without fear of displacement.

Roof Type

There is also a roof-mounted type (D, page 78) which is similar to the vertical ones. It is intended to be fitted to the front of the roof and angled back nearly horizontally. It too has an extra support on the lower section to hold the unit firm and prevent it being flattened down on to the roof and possibly scratching the paintwork. The main difference between this model and the vertical ones is that it is shorter.

The motorised antenna

Although fully retractable antennas are useful in offering protection against vandals, they have one big disadvantage: if the driver wishes to use the radio whilst on a journey and has omitted to raise the antenna before starting, he must stop the vehicle, get out and raise it. If it is of the lockable type, this means fumbling with keys and unlocking it too. If the weather is bad or he is in a no-stopping zone, he will probably decide not to bother and so miss a programme he may have wanted to hear.

The answer to all this is the *motorised antenna* which can be erected and retracted by just switching on from inside the car. This may seem a gimmick, or an unnecessary luxury, but after using it in some of the above situations for a few times one will soon see it as a necessity!

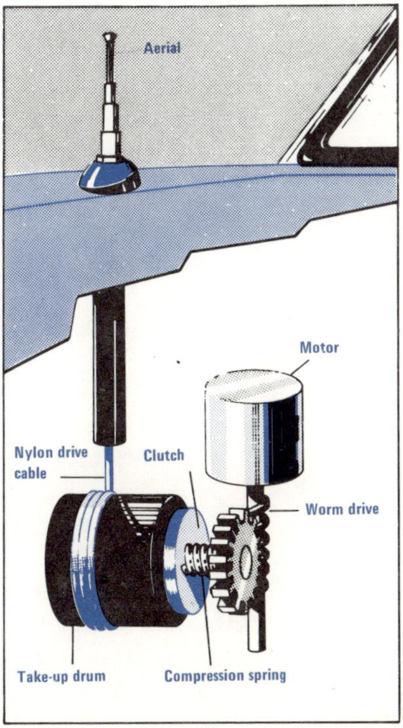

Basic principle of electric motor antenna as used by some makers. A semi-stiff nylon drive cable is stored on a drum which is driven through a clutch inside the drum from a worm drive. The nylon cable pushes the antenna rods up or pulls them down according to the direction of motor rotation. This is changed by altering the polarity of the supply via the up/down switch.

There are many models available from different makers and there are variations in detail, but the basic principle is the same. The antenna itself is a fully retractable telescopic unit, but inside the uppermost section is fitted the end of a stiff but flexible rod about 3 mm (0.1 in) thick made of nylon, polyacetal resin or similar material. It passes down the hollow tubes and is wound for four or five turns around a drum mounted at the base. This drum is driven by either a gear-train or worm-drive from an electric motor.

Exploded view of the clutch. Two balls are pressed against the slots in the clutch plate by the spring-loaded ball retaining plate, thus transmitting the drive. When the antenna is fully extended or retracted, the retaining plate continues to rotate but the steel balls slip from the slots, thus preventing the motor stalling and burning out. The balls click in and out of the slots at each half-revolution until the motor is stopped.

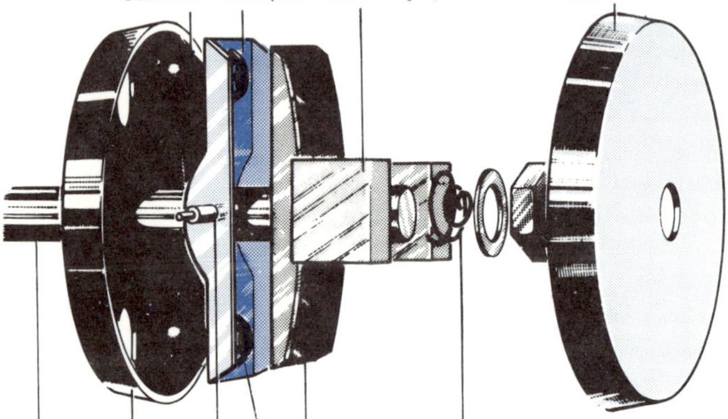

Between the drive and the drum there is a clutch which disengages automatically if the load exceeds a predetermined level. This prevents the motor stalling and burning out when the rods are fully erected or retracted, or should the mechanism become jammed. The most common form of clutch is the ball-bearing type, consisting of two plates thrust together under the force of a compression spring. One plate has a pair of steel balls held in retaining cups, and the other has a couple of matching identations. The balls engage with the indentations, and drive is transmitted during normal running. If the load increases beyond the normal running level, usually between 2.2 and 3.6 kg (5 and 8 lb), the drive plate continues to revolve but the balls slip out of the indentations as the drum remains stationary. The balls slip in and out of the cavities twice each revolution, this being the cause of the clicking noise generated by these units when the rods are fully extended or retracted and the motor is not switched off.

Another drive method used by some makers is to have teeth cut along the length of the flexible rod. Although a drum is used to take up the rod when the antenna is retracted, the drive is applied directly to the rod by a toothed wheel to its teeth.

One factor which must be considered is the space required for mounting the unit. The motor and drive mechanism obviously take up extra space, though as it is usually fitted alongside the lower section of the antenna, the clearance required is in many cases no more than for an ordinary fully retractable. Some models are very short, one being less than 305 mm (1 ft) in length. If clearance is likely to be a problem, look out for one of these.

If the antenna is mounted under the wing, dirt and moisture thrown up by the wheel could affect it and the drive mechanism.

Appearance of complete electric antenna. Motor and drive drums are totally enclosed to prevent contamination from mud thrown up by the road wheels.

As a protection against this the unit is fitted with close-fitting metal covers. Holes are provided at the bottom so that any moisture that might gain entrance can drain out. Should the antenna be in direct line of fire from the wheel, it is recommended that a baffle plate be fitted to deflect anything spun off from it, especially gravel and stones.

In addition to securing the antenna at the usual point where it passes through the bodywork, further support is required at the lower end, but this can in most cases be provided without much difficulty. An alternative arrangement is to mount the motor at a position remote from the antenna. Drive is transmitted to it through a nylon flexible rod which runs in a flexible metal outer sheath similar to a brake cable. This enables an electric antenna to be used where space below the mounting surface is very restricted, with no room for a normal model.

Motorised antenna control

Coming now to control of the motorised antenna, this can be by one of three methods. The simplest is manual switching where a switch fitted on the dash-board is held in one position to erect, and the opposite position to retract, the centre position, to which the switch is biased, being 'off'. The motor has three terminals, one of which is the common ground. When the supply current is applied between ground and one terminal the motor runs in the forward direction, thus raising the antenna, but when it is applied across the other terminal and ground the motor reverses and the antenna retracts. The dash-board control is thus just a simple change-over switch with centre-off position.

The second method is the semi-automatic one. Instead of keeping the switch depressed until the antenna is erected or retracted, it is pressed only to start the required operation, which continues until the rods are fully extended or closed, whereupon the motor automatically switches off. To some extent this is more convenient, but it means that the antenna must always be fully erected and cannot be operated at half-mast, even though this may be all that is necessary when receiving the stronger local stations. The extra unnecessary length may increase wind noise and even slightly increase air drag at high speeds.

The third method controls the antenna by means of the switch on the radio, which means that a connection must be brought out from the radio switch. This system is rather more complicated because the motor must be switched on to raise the antenna not only when the radio is switched on but also when the radio is switched off in order to lower it. Readers who know a little about house electrics will be familiar with the two-way lighting circuit. This works in a similar manner. One change-over switch is fitted in the antenna control unit and is flipped into the opposite position whenever the antenna is fully extended or retracted. The other change-over switch is actuated by a magnetic relay energised from current passing through the radio switch.

What happens is that when the radio is switched on the relay operates and the switch connects the supply to the motor 'up' circuit. The motor runs until the antenna is fully erect, then the antenna switch flips into the 'down' position. However, as the relay switch is still in the 'up' position, the circuit is broken and the motor stops. When the radio is switched off, the relay is de-energised and its switch now reverts to the 'at rest' position which connects the supply to the 'down' circuit. As the antenna switch is also in the 'down' position, current flows to the motor and the antenna retracts. On reaching the fully closed state, the antenna switch is again flipped, breaking the circuit and putting it in the 'up' position ready for the next switch-on.

The current through the relay is quite small, so the radio switch is not overloaded and the radio fuse need not be increased in value. Motor current is supplied separately and so should have its own fuse. Values vary considerably from make to make, and can be as low as 1.5 A or as high as 7 A; 6 V models are even higher and can take up to 14 A. These currents of course are taken only for a few seconds at a time, so will not constitute a serious drain on the battery.

Antenna finishes

Something that should be considered with all types of antenna is the finish. Chrome-on-brass or stainless steel is the preferred material. Inferior finishes may look just as good when new, but quickly deteriorate. Tarnish or corrosion not only looks bad but can cause a resistance to develop between the telescopic sections. Thus the upper sections can be virtually isolated from the bottom ones and might just as well not be there at all. Even with the better finishes, the rods should be kept clean to ensure good mutual contact, and a smear of oil will help.

Antenna leads

There would appear to be nothing very special required of the wire used to connect the antenna to the radio. It would seem that a straightforward length of insulated wire would do the trick. If this were used, however, it would pick up a great deal of interference from the car electrical system, and the resulting noise would probably drown out the radio programme completely. It is necessary, therefore, to use shielded wire to protect the lead from stray interference fields. The shielding is provided by completely encasing the insulated wire with copper-wire braiding. This braiding is connected to the car body at the antenna and also to the outer contact of the antenna plug. Any picked-up interference currents are thereby grounded.

However, using shielded cable introduces a complication. The inside surface of the braiding provides a comparatively large conducting area in close proximity to the inner conductor, and this produces the property of *capacitance*. The cable is therefore in effect a capacitor or condenser, which is connected in parallel with the radio input.

Now radio tuned circuits have as their constituent components coils and capacitors, and varying the value of either the coil inductance or condenser capacitance alters the tuning of the circuit. It follows that connecting another source of capacitance across the input circuit increases the total amount and so changes the tuning. As there are a number of tuned circuits in the radio, all accurately aligned to work together, the detuning of the input circuit by the antenna cable will obviously produce misalignment and reduction of peak performance. The actual effect is loss of volume on weaker stations, and increase in noise and interference.

The set makers anticipate this extra capacitance and so allow for it in the design of the input stage, but as cable capacitance is proportional to cable length and lengths differ, as well as capacitance per unit of length varying from one sample to another, they cannot accurately allow for all possible car installations. The solution to the problem lies in the provision of a small variable trimming capacitor on the radio which can be adjusted to compensate for various cable capacitance values. It is usually adjusted through a hole in the radio casing by means of a screwdriver, with the radio connected to the antenna it is to be used with.

A problem can arise when the antenna is mounted at some distance from the radio, such as at the rear of the car. The supplied cable will almost certainly be too short and so will need to be extended. The best way of doing this is to buy a ready-made extension lead and just plug it on to the existing one.

Failing this, extra cable can be joined on and a plug fitted to the new end. This means soldering the plug, and preferably the joint, although a terminal block can be used for the latter. The cable chosen should not be of the shielded audio-lead type which has a very high capacitance, nor the television antenna coaxial sort. It should be the special low capacitance cable made for the purpose.

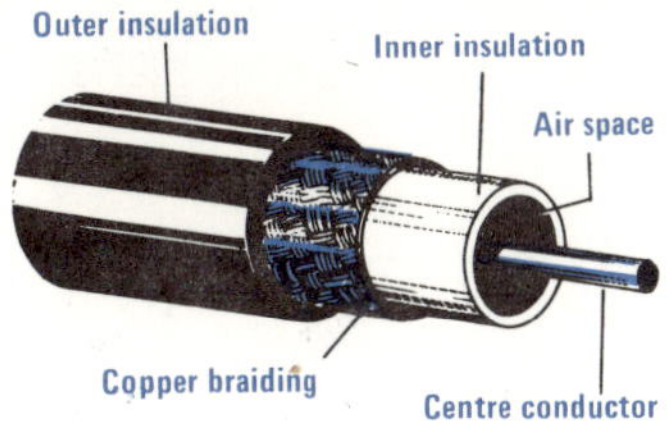

Low capacitance antenna cable. The centre conductor is a fine wire running inside a much larger insulated tube. Thus the large spacing between it and the copper braiding and small facing area give reduced capacitance value.

This can be identified by the very fine gauge centre wire and the semi-rigid insulation tubing which is of much greater diameter than the wire, thus producing an air gap between them. The surface area of the wire and the proximity to the shielding is thus kept to a minimum. If other cable is used, the capacitance may be too great for the antenna trimmer to compensate for. Capacitances of less than 100 pF are necessary – anything much in excess of this will most likely be beyond the trimmer's range.

9. Interference suppression

Interference is any sort of noise that gets into the radio circuits from an outside source. In the case of car radios it is nearly always generated by the car electrical circuits. If an electrical circuit carrying a fair current is switched on or off, a spark will probably appear across the switch contacts. Energy is released and this produces surges backwards and forwards in the associated circuits until the energy is dissipated. The effect is similar to a stone being dropped in a small pond, when ripples travel out to the edge, back to the centre and out again until they die down. This happens very rapidly in the electrical circuit and the oscillations produce radio waves over a wide but random range of frequencies. At least some of them will fall within the range of the receiver tuned circuits and so they pass through them to emerge from the loudspeaker as noise. Certain types of interference contain a higher proportion of some frequencies than others, which means that some radio bands will be more affected than others.

Interference sources in a car

There are many spark sources in a modern automobile, the ignition being the most prolific, as of course its sole purpose is to produce sparks. There are others, though: the generator, voltage regulator and many accessories all do their bit, so it can be seen that we have a real problem!

One fact must be understood, and that is that although the spark itself is the source of interference and radiates it to some extent, more interference is usually radiated by the associated leads, because it is through these that the oscillations are taking place. They act therefore as a form of transmitting antenna. This radiation can be picked up by other wires not directly connected to the interfering circuit, or even nearby metallic objects such as pipes, body panels or bowden cables, and then re-radiated. This secondary radiation can also find its way into the radio circuits.

There is also another form of interference. Large masses of metal such as body panels and road wheels can build up electrostatic charges during periods of dry weather. This is similar to the charges that build up on discs which attract dust. Periodically these charges jump over to adjacent metal-work, producing the same sort of interference-generating surges that we get from switching.

Direct current (d.c.) electric motors have a device known as a *commutator* which alternately switches-in rotor windings as the rotor turns. This is essential to its functioning, but as may be imagined it generates considerable interference in the form of a continuous crackling noise. The same is true of a generator which is basically just a motor working the opposite way. So, in addition to the ignition system and regulator, all motor-operated functions in the car produce their own interference. Some of these are the gas pump, water pump, windshield wipers, heater motor and starter motor. The starter takes a very heavy current and so its surges generate severe interference. However, as it is only used for a few seconds to start the car engine it can be disregarded.

Both primary and secondary circuits of the ignition system produce interference, so we can have interference radiation, for example, from the ignition switch wiring which is in the primary circuit. This is so whether the switching is by conventional switch contacts (contact breaker ignition) or by a transistor (transistor operated ignition); switching surges are still present whenever a current is suddenly interrupted.

Interference suppression

There are several methods of suppressing interference. One is by inserting *resistance* in the circuit between the interference source and the radiating wiring (A). This reduces the oscillations but does not completely eliminate them. As any part of the wiring between the source and the resistor will radiate, it must be fixed as close to the source as possible. The snag with resistance suppression is that it will produce power loss in the circuit if there is any appreciable current flowing. It can therefore only be used in low current circuits, and even then the resistance value must be carefully chosen to give sufficient suppression with minimal power loss.

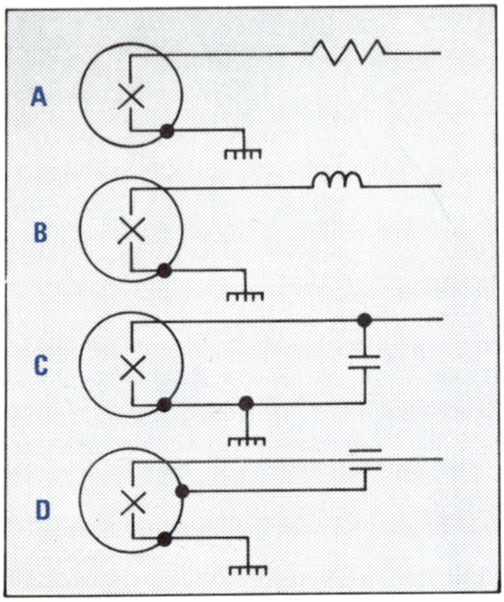

Various types of suppressor. A Resistor suppressor minimises flow of interference energy to connecting lead. B Inductive suppressor offers a high impedance to high frequency oscillations but a minimum to d.c. currents; especially useful for v.h.f. interference suppression. C Capacitive shunt serves as a 'short circuit' to h.f. currents. D Coaxial capacitor wired close to terminal eliminates radiation by the connecting lead of ordinary capacitor; useful for v.h.f. suppression.

Another form of suppression is the *choke* or *coil* (B). A choke has the property of offering a high impedance to high frequency currents, but zero impedance to d.c. It can be seen that this is ideal for suppression as it is the high frequency interfering surges that we want to impede, but with no power loss at d.c. A coil does impede a.c. currents, so some power loss will be suffered if used in a.c. circuits such as an alternator. The effectiveness of choke suppressors can be limited by the self-capacity between the windings. Capacitance has the opposite effect to a coil, offering a high impedance to d.c. but low impedance to high frequency currents. Thus the self-capacitance will to some extent cancel the effect of the choke suppression. With a well designed unit, though, the capacitance is small compared to the inductance and the cancelling effect minimal. The moral here is to buy only good quality suppressors.

The third suppression method is the *capacitor suppressor* (C). As noted before, a capacitor readily passes high frequency currents, but blocks d.c. It can then be wired not in series, but shunted across the interference source. Thus the interfering surges will flow through the capacitor and be bypassed.

This too must be wired as close to the source as possible because any wiring between it and the source will radiate. The capacitor lead itself will radiate, but as this is only a couple of inches the radiation is insufficient to cause trouble on MW. On short waves, and especially on v.h.f., the radiation from even this short length may prove troublesome. To avoid this, special coaxial or lead-through capacitors are used (D). These are connected in series with the lead-out wire from the source, close up to the terminal. The capacitor is not itself in series with this lead, as that would obviously block the d.c., but the capacitor surrounds the lead which passes through the middle of it and is connected internally to one of the condenser plates. The outer plate must be connected to the casing of the interference source directly, and not to nearby bodywork as this would cause the h.f. currents through the capacitor to flow through the bodywork and so radiate from it. So then, the capacitor must be connected right across the source as closely as possible, and take in no external path.

The construction of many condensers, being wound up from a roll of metal foil, gives them the property of inductance just like a choke. This being the opposite of capacitance partly counteracts the suppressor effect. Although not great it can be enough to make trouble at v.h.f. and the higher short wavebands. Capacitive suppressors should therefore be non-inductive, and most good quality ones are in fact of this type.

Static suppression

Static between panels and secondary radiation can be dealt with by bonding adjacent panels that may not be in perfect electrical contact, with copper braiding. It may be necessary to so bond the motor block to the bulkhead, or the differential housing to the body. The braiding should be thick and should be tinned with solder at the points where it is bolted to the panels. If this is not done, electrolytic action may start in the presence of a slightly acid moisture between the dissimilar metals of copper and steel. This would result in corrosion at that point, which if nothing else would destroy the electrical contact and so render the bond useless.

Shielding

The shielding of interference sources, radiating wiring, or radio circuits by grounded metal-work greatly reduces coupling between them, hence interference pickup. Thus the radio has a metal case and the antenna cable is shielded with copper braid. The spark plugs are buried in the motor block and so are effectively shielded, but their high tension leads are not and neither is the distributor. Siting the antenna on the opposite side of the car to the plugs puts the block between it and them and the distributor, and so affords a measure of shielding. The car body and hood lid also shield the antenna from ignition interference.

A particular problem arises in the case of glass-fibre bodies. They offer no shielding at all, being non-metallic, and there can be serious interference. The only solution is to line the motor compartment with metal foil. This should be done to all accessible places including the underside of the hood. All sections should be linked with braiding, and a strap should be taken from the nearest foil to the motor block.

Preliminary measures

All car radio installations should include three measures. One is a suppressor in the high tension (h.t.) lead from coil to distributor. If a resistive brush to the rotor arm is fitted or a resistive lead composed of carbon granules, no further suppressor is needed here. Otherwise a 10 kΩ series suppressor should be fitted as close to the distributor as possible.

Next, the coil should be fitted with a 2 μF capacitor suppressor connected from the low tension (l.t.) lead to the ignition switch to the bracket securing the coil. This should not be fitted to transistor or capacitive discharge systems. The dynamo, where used, should have a 0.5 μF capacitor wired from the D terminal to the frame. No capacitor should be connected to the F terminal; if in any doubt, the D terminal has the thicker wire connected to it because this carries the whole output current from the dynamo. The F terminal wire carries only the field energising current.

Having suppressed these points, interference should be reduced to a tolerable level. This means it should be completely absent when receiving strong stations, and only just audible in the background with weak ones. If it isn't, we have to trace the source of the trouble, but before doing this check that the antenna is in good order, especially if it has been fitted for some time. Make sure that all sections are in good contact with each other — an ohmmeter

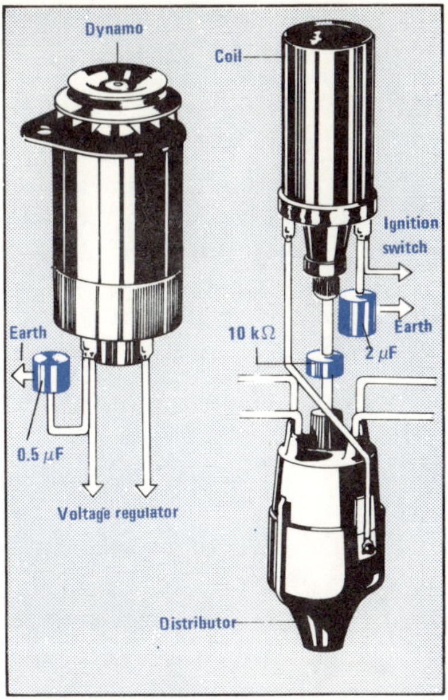

Routine suppression for reception of a.m. radio. A 10k resistive suppressor in the h.t. lead to the distributor, a 2 μF capacitive type wired close to the coil primary terminal and a 0.5 μF to the D terminal of the dynamo.

or circuit tester applied across the top and bottom section will show this. Check that there is continuity between the antenna and the radio plug, and in particular that there is no leakage between the antenna and the bodywork. Moisture across the insulator sometimes produces a leakage, especially if there is a deposit of dirt and mud over it. Check also that the cable shielding is well grounded; it should show a very low reading on an ohmmeter, just a fraction of an ohm. Trouble with any of these points will decrease the signal and increase interference.

The ignition system too should be in good condition. Worn plugs or rotor arm or any loose connection will produce excessive interference.

Tracking down interference sources

Tracking down the interference source is not too difficult as the symptoms can be easily identified. Ignition system produces a sharp ticking noise that varies with engine speed and ceases immediately the engine is switched off. The dynamo gives a whining noise which increases in pitch as the motor is revved, but it is quite different from the ignition noise. It dies down when the motor is switched off and this helps to distinguish it from ignition interference. An alternator has a similar effect.

The voltage regulator is quite distinctive, giving what could be called a chattering effect. It is heard only when the revs are high and ceases when they fall below a certain speed. Another source is the instrument stabiliser. This stabilises the supply voltage from the battery so that instruments that are voltage operated are not subject to inaccuracy due to voltage fluctuations; it is not though found on all vehicles. It uses a bimetallic strip and small heating element which continually opens and closes a pair of contacts. This happens so rapidly that the damped instrument gauges behave as though a lower but constant voltage is being applied to them. The noise it generates sounds like a continuous sizzle which appears when the ignition is turned on irrespective of whether the motor is running or not. It is often made worse by vibration, as sometimes can be discovered if the instrument panel is struck sharply with the hand. The presence of one of these devices can be determined by observing the fuel gauge when the ignition is turned on. If it registers immediately there is no stabiliser, but if it takes a few seconds, one is present.

Interference from other ancillary equipment can be readily identified by switching the particular items on and off. Heater motor, wipers and direction indicators are the main culprits. An electric clock will give a ticking noise just like a mechanical one over the radio, and of course it is on all the time when everything else is switched off.

Static due to charges in the body panels is usually only heard when the vehicle is in motion, and is best heard when everything else is off. It can be identified by coasting down a hill with the engine stopped. A similar effect may be had from a poor electrical circuit contact, but this will only appear when the particular circuit is switched on; it should also show itself by erratic behaviour of the item supplied by the circuit, or below standard operation. An example is that of a dim lamp which is most likely due to a poor ground connection and which also could generate interference.

Further suppression measures

Having identified the source of the interference, further measures can be taken to suppress it. There may of course be more than one culprit, and each one will have to be dealt with separately.

All possible suppression points, rarely necessary except for f.m. (v.h.f.) reception. The extra points are: 5k each in plug leads, non-capacitive filter in dynamo F lead, and 2 μF coaxial to the voltage regulator B terminal. An inductive filter can be wired in series with the power lead to the radio, and each electric motor such as wiper, heater, window winders, etc., should be fitted with a 0.5 μF suppressor.

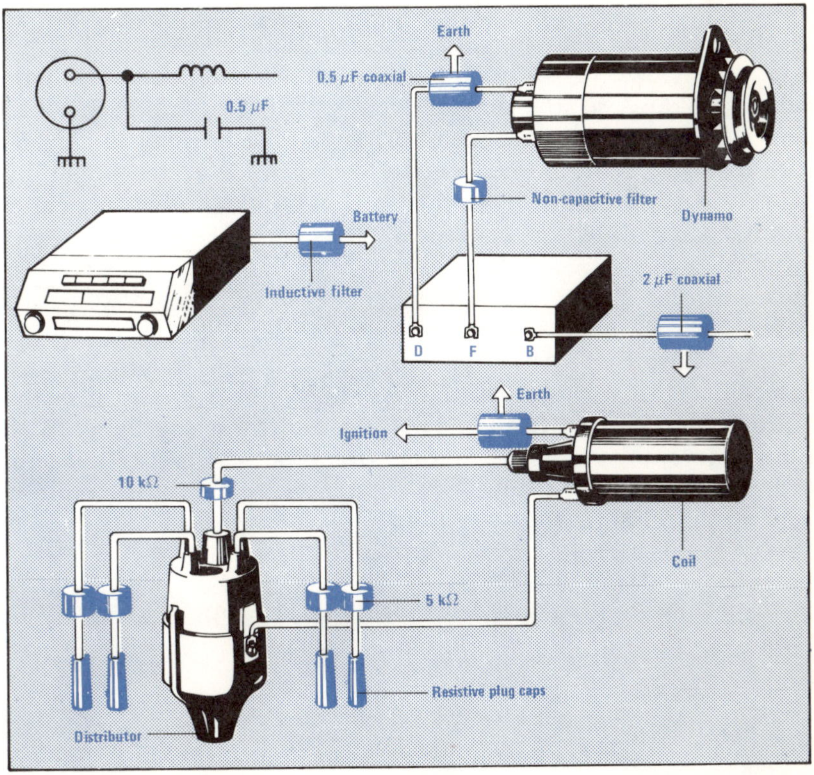

Persistent ignition interference may yield to extra plug suppression. An extra 5 kΩ suppressor in each of the leads wired close to the suppressor or, if available, the type that plugs into the output spouts of the distributor may do the trick. Plug suppressors are usually resistive, but inductive ones (choke type) are most effective if they can be obtained, especially for v.h.f. interference.

Continued trouble from ignition may be due to radiation from the ignition switch wiring. A 0.5 μF capacitive suppressor connected from the ignition switch to the body should take care of this. A grounding braid taken from the coil housing directly to the motor block will eliminate high tension return currents passing around the bodywork and possibly radiating interference, so this could be tried in persistent cases. The ordinary 2 μF coil suppressor could be substituted for a coaxial lead-through type. In some cases taking the power lead to the radio straight to the battery instead of through the car junction box has been known to improve matters. Inductive filters are usually included in the radio power circuits, but a filter could be tried back at the junction box end of the lead.

Loudspeaker wiring can pick up interference and pass it inside the radio shielding to the input circuits. Try re-routing the leads around the region of the dash-board, or even running in shielded wire instead of the normal twin. Do not overlook the possibility of secondary radiation from the hood; a bonding braid from hood to bulkhead should take care of this.

In the case of dynamo interference, fit a 0.5 μF suppressor to the D terminal of the voltage regulator. The lead to this terminal is already suppressed at the dynamo end, but an extra one here will ensure that this wire is not radiating. The suppressor at the dynamo could be substituted for a coaxial one and the value increased to 1.0 μF. In stubborn cases an inductive filter can be tried in series with the F terminal lead, but no capacitive suppressor should be used on this circuit.

Voltage regulator interference can usually be eliminated by wiring a 2 μF suppressor from the B terminal to the body.

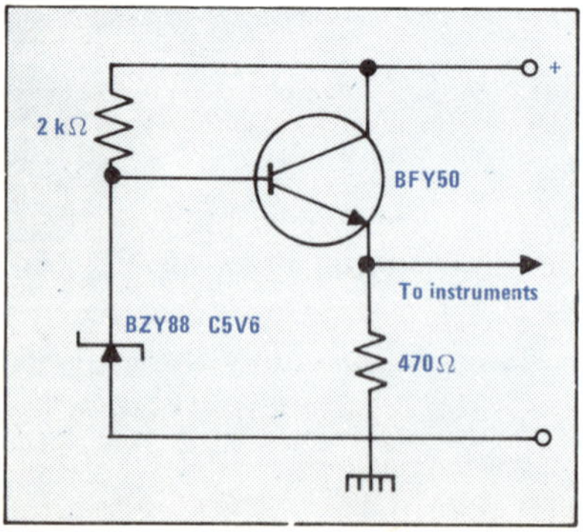

A circuit for a transistorised instrument voltage stabiliser to take place of the electromechanical type which can generate interference. Only two resistors, a transistor and a zener diode are needed.

Trouble from the voltage stabiliser can in some cases be cured by wiring a 0.5 μF suppressor from the terminal that feeds the instruments to the body. A series choke may be required as well. If this is not successful, a transistor circuit can be used in place of the mechanical stabiliser. It uses a transistor, zener diode and two resistors. As it supplies current at a steady regulated voltage, it does not switch on and off as does the mechanical stabiliser, hence does not generate interference. No particular layout is required, but a heat sink, a piece of heavy gauge metal in intimate contact with the transistor, should be provided to cool it (page 96).

Most ancillary equipment interference can be dealt with by wiring a 0.5 μF across the offending item. If this does not do the trick a series inductive suppressor will be needed as well. There are suppressors available that combine a capacitor and choke in the same unit, and they are particularly effective and convenient to use for this purpose.

Static

Now we come to the final source of interference, that of static. Unlike electrical sources, static can be difficult to trace because there are few clues to pinpoint the actual cause. About the only way of tackling the problem is the lengthy one of trial and error. Try bonding the wings and hood, and if the antenna is in the rear, the trunk lid. The exhaust, steering column and differential housing are other possibilities.

Tyre static is another source. Friction can build up a charge between ground and car bodywork that leaks only when it has reached high potentials. This can cause not only radio static but even an electric shock when someone dismounts. A dangling chain can help to discharge static to ground, but its effect is rather uncertain. The most effective cure is to provide a conductive path from the tyre tread to the car body. Painting the inside wall of the tyre with lead paint or a rubber conductive paint is one of the best ways of doing this. It is not necessary to paint the whole wall – several wide strips should do, but take the paint right up to the tread.

The wheels themselves may not be in good contact with the body through the wheel bearings. Packing them with graphite grease should make them sufficiently conductive. Alternatively, special springs made for the purpose can be fitted inside the hub caps. These form a bridge between the hub cap and the axle.

In the case of the rear wheels it may be found that the back axle is not in electrical contact with the body, and if this is so it will be necessary to fit a copper braid strap from the axle to the body. Static usually appears during long dry spells when any dampness that may have provided a leakage path has dried out. It is more likely to be troublesome, therefore, in drier areas.

For f.m. or v.h.f. installations the three routine suppression points will be insufficient and some of the measures described for stubborn cases will almost certainly be necessary.

10. Installing the system

When preparing to fix up the various units in the car, the first thing is to make sure that all needed items are to hand. There is nothing more frustrating than to start a job such as this and then find that some vital part is missing, especially if the task is being done on a day when the stores are shut and there is no chance of obtaining it.

Mounting the radio

The radio usually comes complete with basic mounting kit including screws, rear support bracket (not always used for a radio but essential for a playing unit), fascia plate which fits to the front and overlaps the car fascia panel, nuts for the control spindle bushes which secure the fascia plate to the unit and the unit to the car fascia if so mounted, washers of different thicknesses to fit under the nuts, and/or decorative plastics washers to take up any gap between the fascia and the knobs. The washers used should be of appropriate thickness and the others discarded. Sundry screws, self-tapping and otherwise, are also normally included. There will also be a power lead with fuse. Check off these items to make sure nothing has been left out.

Some makers have special kits available as optional extras. They are designed for particular model cars, and consist of fascia plates and knobs in the same style as the car instrument panel. This enables a match to be made which gives the unit a custom-built appearance.

In addition to the kit, check that materials are to hand for wiring and mounting the loudspeakers and antenna. Do not forget the speaker wiring; have enough to run to the position chosen, and allow some extra – often a run takes more than anticipated because of some unexpected detour that must be made. If a stereo outfit is being installed, the wiring must be of the two-colour type (a different colour for each conductor of the pair) so that correct phasing can be made.

Now we have to decide where to mount the radio or player unit. There are three principal positions: through the aperture in the fascia, under the front parcel shelf or dash and in a console over the gear box housing.

Fascia mounting

The most common, and usually the most convenient, is through the aperture in the fascia which is covered by a dummy panel when no radio is fitted. All that is necessary here is to remove the panel, but it is advisable to keep it safely in case it is required to remove

Radio installation in the dash-board. Side brackets are usually provided all ready for the supporting of the radio. They have horizontal slots so that the radio can be positioned. A rear bracket or perforated strip secures the back of the unit to the bulkhead. Fascia plate, station scale and knobs are fitted from the front after the radio is in place.

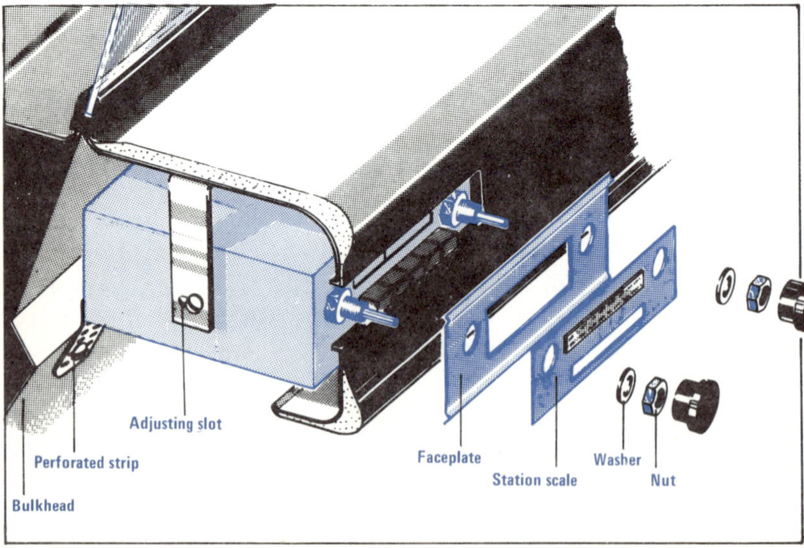

the unit later when the car is traded in. In some cases, holes may have to be drilled in the instrument panel to accommodate the control spindles, but it is essential before doing so to double check that there is room behind the panel at that point for the unit before you start drilling. There should be few problems with modern radios as they are compact, but some player units may produce a tight squeeze.

Actual fitting differs from one model to another and from one car to another. The larger radios and most players need to be secured both at the front and the rear. A common method is by means of bolts at the side of the unit which pass through slotted holes in a pair of brackets already fitted behind the aperture in the fascia. These are usually bendable, and this flexibility together with the slotted holes enable units of different dimensions to be fitted. Some radios are secured at the front solely by their control bushes. These are easier to fit, but also easier to steal, and theft of car radios is unfortunately a growing risk.

Rear fixing serves to steady up the unit, and take the strain off the front fascia mounting as well as providing some deterrent for would-be thieves. This is done by means of a length of perforated mild steel. It is fitted to the back of the unit by means of the supplied bolt, then bent back to a convenient point on the bulkhead where a hole is drilled and the strip secured through one of the perforations by means of a self-tapping screw. In some cars a threaded bush and bolt will be found on the bulkhead for this purpose, thus saving the drilling and tapping. In most cases the strip will be longer than actually needed, so the excess can be cut off before fitting permanently.

Underslinging

The second position is under the dash or parcel shelf. The unit can of course be fitted on the parcel shelf, but this would take up valuable room, and however big they are, shelves never seem to have enough room to accommodate all the bits-and-pieces that seem to accumulate! Underslinging is not quite so elegant as the through-the-panel position, and it also means that the controls are lower and not so easy to reach when driving. Before deciding on this position, make sure that they can in fact be operated without causing a hazard. Ensure too in the case of tape players that there is sufficient room in front of the unit, otherwise it may be found that the gear lever makes it impossible to insert a cartridge! Many players are too high to mount in the aperture, and so must be fitted elsewhere.

A bracket, called a *gimbal*, is one means of underslinging a unit (A). It fits to the sides and passes over the top of it, and must first be fitted to the underside surface by self-tapping screws. If fitted to a shelf, see that the shelf is solid enough to support the weight – a good many are not!

Various under-shelf mounting arrangements. A Gimbal running over the top of the unit. B L-shaped side brackets. C Box-type fascia plate. This and the gimbal must be fitted to the underside of the shelf first before the radio can be mounted to it.

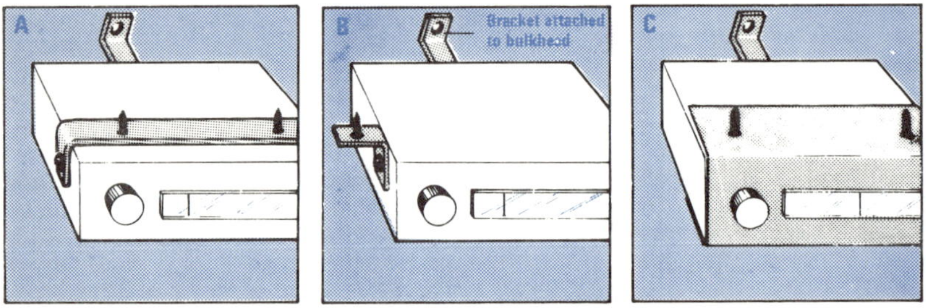

The gimbal allows the angle of the instrument to be adjusted as required. If the underside surface is sloping, this can be compensated for, or it may be desirable to tilt the unit upward slightly to improve visibility and make operation of the controls easier. Tape players, though, should not be inclined too far from the horizontal unless the maker states that this is permissible, as it could cause speed irregularities and other mechanical malfunctions. About 20° is the maximum tilt allowed.

Instead of the gimbal, two L brackets can be used to undersling the unit (B), or some makers supply a special faceplate made in box construction with an open back (C). This is first screwed into place, then the radio can be introduced from the rear and secured by the control spindle bushes and nuts. Yet another type of fixing uses a metal tray with high sides and drilled front. This is really an extension of the box faceplate.

For radios of the non-stereo variety it is sometimes convenient to mount the unit in one piece with the loudspeaker. Some radio manufacturers make boxes for this purpose with space for the radio and also the speaker. They are underslung, with the speaker facing downward toward the floor, either directly, or more usually with a slight forward angle. If there is room between the underside of the dash and the transmission hump, a box can be chosen that allows the speaker to face outward. This is better acoustically because high frequency tones tend to travel in straight lines and are easily absorbed by soft surfaces such as carpeting. A downward facing speaker will therefore direct the high frequencies into the carpet and the reproduction will be dull sounding and lacking in clarity. Front-facing units usually use elliptical speakers instead of round ones, as the depth can thereby be kept to a minimum.

Console fitting

The third method is that of a console which fits over the gear box and rises to meet the bottom of the instrument panel. Many cars have a console already built in with heater and other controls fitted, and space is provided for installing a radio or tape player. Mounting here is the same as for the in-dash fitting. Some makers have a range of consoles designed for particular popular model cars. Both radio and speaker can be easily fitted and the console

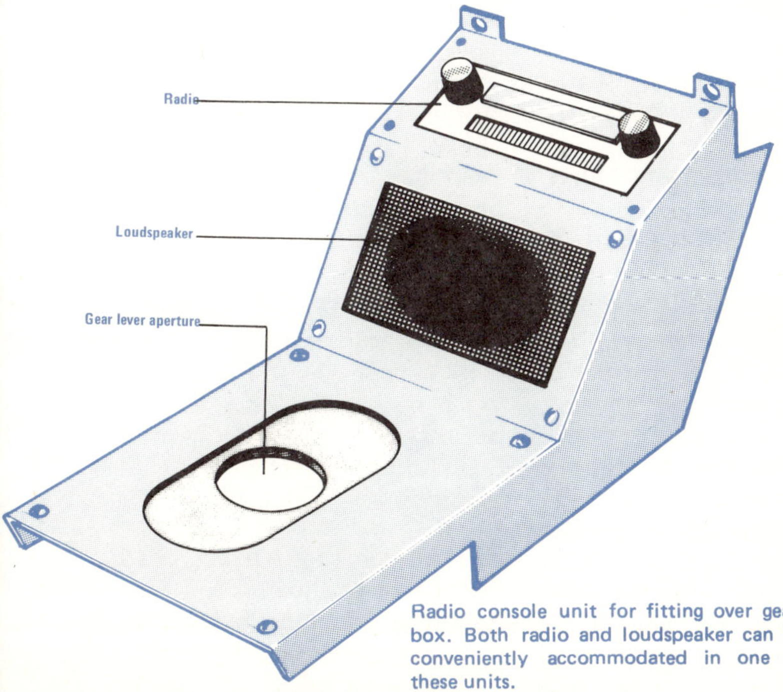

Radio console unit for fitting over gearbox. Both radio and loudspeaker can be conveniently accommodated in one of these units.

is secured by a few self-tapping screws to appropriate spots on the bodywork. A feature of these consoles is the tray that surrounds the gear lever, which utilises otherwise wasted space to store maps, guides, sunglasses and other small items.

Loudspeakers

Having decided on the location of the radio or player, the next job is to work out where to fix the speakers. It would be wise not to start fitting the player until the speaker locations have been decided on, as the situation may call for some modification of the player mounting.

With a non-stereo radio, only one speaker will be involved (unless one is wanted at the rear as well), so this should be fitted in a central position, if possible, to give good results for both driver and passenger. This requirement will be met by the combined under-dash radio/speaker mountings and gearbox consoles already described. Other than these, an under-dash fitting using the most convenient of the various speaker housings will usually be the most satisfactory arrangement. Try to get it facing outward rather than downward to preserve a crisp reproduction.

Of course, more than one speaker can be used for a radio, either an extra one for the back or ones on either side. This will not give stereo radio but can give improved results. These can be controlled by switches and faders as described in Chapter 2. It may be worth giving some thought to the possibility that a radio-only system may be changed in the future to that of a radio/stereo outfit. If two speakers are already installed, they will be ready for the change later with the minimum of extra work. Any centrally located speaker would be of little use in such a case and would have to be removed.

Fascia-top fitting

Pod-type speakers can be fitted to the upper surface of the fascia, underneath the windshield. This position, being nearly at the listener's head level, is probably one of the best for preserving the high frequencies. If a stereo pair are directed *forward* so that the sound is bounced off the windshield as described in Chapter 2, a good stereo image will be obtained for both driver and passenger.

Driver's vision of the road ahead must not of course be obscured in any way, so this point will have to be carefully considered before this position is adopted. Wiring to such front fascia speakers can be untidy unless it is taken through a hole drilled in the fascia top. The snag is that it is not easy to drill because of the windshield rake. A flexible drill extension might help here. A rubber grommet should always be used when passing wire through a hole in metal as it protects the wire from the metal edge and also looks more presentable.

Door and kick panel fitting

Another possible mounting position is the kick panel or scuttle on each side of the car; and one that is becoming increasingly popular is the door panels. A speaker so mounted is well baffled and should give plenty of bass, although being low down and partly obscured by the legs of driver and passenger, the treble is rather poor. The extreme side positioning also gives a bad stereo image.

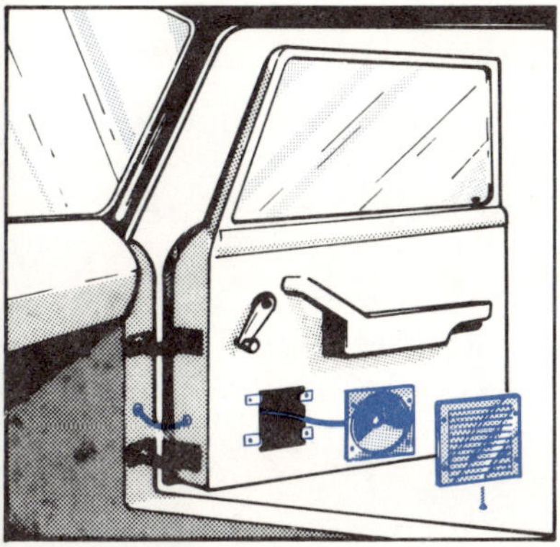

Door fitting of loudspeakers. Cable holes in door and pillar must be fitted with rubber grommets and staggered to avoid cable chafing and fatigue. Speaker must be mounted with drip shield uppermost and care should be taken that it does not foul any window winding mechanism *before* cutting the panel hole.

A suggestion for hi-fi enthusiasts is to use two speakers per channel, one for the bass and the other for the treble; a crossover network which can be bought at any hi-fi component store can be used to separate the lows from the highs. The bass speaker can be mounted in the door, while the treble one can be fitted to the fascia top directed towards the windshield. The spatial separation between the low and the high speaker is far from ideal theoretically, but a good stereo effect (which depends mainly on the treble frequencies) can be obtained with a good bass response too. Wiring for the door speaker should be arranged as shown, with the holes in the door and door frame staggered and a loop left to prevent cable fatigue and breaking.

There are several points to watch when fitting a loudspeaker in a door panel. It must be remembered that rain water can run down the inside of the door from the window to escape from the drain holes at the bottom. Speakers intended for this position have drip shields over the upper part. Speakers without such protection would soon have their cones reduced to a soggy mess after a heavy storm or a car-wash. Not only must the speaker have such a shield but it must be fitted with the shield uppermost over the unit.

When marking out the position in the panel, check that the speaker will not be obscured by the seat when the door is closed. Also see that the winding handle does not foul against the speaker grille or that the back of the speaker will not obstruct any part of the winding mechanism for opening the window inside the door. If the hole cut for the speaker is too large there may be nothing for the fixing screws to hold to, so care must be taken to get the size right. The screws will be secured by spring clips at the back of the panel.

For quad reproduction, the front two speakers can be mounted just the same as for two-channel stereo. The rear ones though should be high and not in the rear door panels. The back parcel shelf is the best place, and pod-type surface speakers can be fitted there, or units can be flush-mounted into the shelf. Pod speakers can obscure the rear vision, although this is probably not too serious, especially if wing mirrors are used in addition to the interior one. While the flush-mounted speakers point upward, the rake of the rear window will bounce the sound forward so there is no problem there.

Antennas

There is quite a choice of locations for the antenna, either of the front wings, side pillars, back wings or even the roof. Most antennas have enough cable only for front mounting so any rear position will mean extra cable. Aim to keep the cable within reasonable lengths, otherwise the capacitance will be too great for the antenna trimmer in the radio to compensate for. To minimise interference pickup the antenna should be fitted on the opposite side to the motor distributor and plugs. With rear-engine cars this gives a choice of sides at the front, but the nearside is preferable in the case of f.m. radios, to give a little extra shielding by the car body from interference from traffic going in the opposite direction. For the same reason rear mounting is best for f.m. with cars having the engine at the front.

Whatever the chosen position, before doing any drilling, check and double check that there is sufficient clearance underneath for the lower part of the antenna, especially if it is a retracting type. Make sure too that there is access beneath the selected spot sufficient to enable the antenna to be introduced into the hole from the underside. Some non-retractables, roof and pillar-mounted types are not fitted from the inside but do need an internal exit space for the cable to pass. Some car bodies have box sections with no access to the inside, so look out for anything like this before drilling.

Motor antennas and some other retractables have a bracket on the lower end to steady them, and this must be fitted to the bodywork by self-tapping screws. A suitable position for this must also be sought before starting to drill.

Having made sure of these points, the hole can be drilled in the required spot. It is usually 25 mm (1 in) in diameter, but some are smaller – check the antenna maker's instructions. It can be rather an anxious moment because a mistake or a slip can have disastrous results. To ensure that the hole is made in the right spot and that the drill does not wander, as it can do on a smooth metal surface, first punch a dimple with a hammer and punch or, if a punch is not available, a large nail. Hold it near the bottom so that it will not slip when struck. Now drill a pilot hole of a smaller diameter, then finally the full-sized hole.

The casing of the antenna to which the braiding of the cable is grounded must be in first-class contact with the car body, and to accomplish this many antennas include a serrated-edged saddle or bar which bites into the metal when the securing nut is tightened. To make quite sure of a good contact though, especially with models not having this device, it is wise to clean around the underside of the hole right down to the bright metal before fitting the antenna. Then assemble according to the maker's instructions and bolt up tight.

Wiring the speakers

With everything installed in place, or at least partly so, we can start the wiring up. The speakers can be connected first. Ordinary twin wiring can be used such as fitted to light domestic appliances, but it should be of different colours if the system is in stereo. As with any other form of wiring, make good connections and avoid anything that could short either one wire to the other or one wire to the car body. Wire-whiskers at the connection point are the most likely source of trouble, so these should be carefully avoided. When passing wire through holes in the metal-work always use a grommet to avoid chafing the wire.

Wiring to front-mounted speakers is usually no problem as the run is short and concealed behind the fascia. With rear-mounted speakers the wiring must be run along the length of the car. One method is to drill a hole through the floor at the front and back and run the wiring underneath. To avoid hanging in a loop it must then be supported at several points along its length, and it is vulnerable to damage by stones, jacks and other hazards. Usually it is better to have the wiring inside the body. The path will depend on the model car and the most convenient place to run it, but when laying it under the carpet, keep it either well to the side or close to the central transmission hump. In this way it will not be stepped on by the rear passengers, and not be unduly disturbed when the carpets are taken up for cleaning. The wire can be run up behind the carpet on the bulkhead and from there behind the dash to the radio or player. Keep it well away from the pedals.

Where more than one speaker is used, whether for stereo or two-speaker mono, they must be connected the same way so that the cones move backwards and forwards in unison and one is not going backward while the other is moving forward. This is known as keeping the speakers *in phase* and if not observed can lead to poor stereo image and a thin-sounding reproduction. It requires colour-coded wiring and connecting like terminals on speakers to like terminals on the player or radio. It does not matter whether switches or faders are interposed in the wiring, the speakers must still be connected in phase.

Antenna cable

In most cases the antenna cable is already fitted to the antenna and
 supplied with a plug so there are no connection difficulties.

If it is not long enough then the easiest and best solution is to buy an extension length complete with plug and socket so that it is just plugged on to the existing cable. When running an extension ensure that there is no strain on the join or that it is not in a position where it could come apart. Alternatively, the plug can be cut off, an extra length fitted by means of a terminal block, and a new plug fitted to the end. It will be easier to do this than trying to fit the old plug which was soldered to the cable. Make sure too that only low-capacitance antenna cable is used and that the shields and centre conductors are not reversed but are connected to each other at the joint.

Try to avoid running the antenna cable through the engine compartment where there are strong interference fields; also keep it separate from other wiring, especially ignition wiring.

Power wiring

Finally the power connection can be made. Before doing so though, double check that the polarity of the unit is correct for the vehicle. It is quite common for an owner to fit a unit, perhaps taken from another car, without considering this essential point, and the result can be burnt-out transistors. Many radios have a dual-polarity switch so that the unit can be simply set to whichever polarity is required, either negative or positive ground; others have internal wired links that must be resoldered in a different position. Information regarding this is usually marked on the case. Many radios though and most player units are made for only one polarity, negative ground, because nearly all cars made today have their negative battery poles grounded. When installing in an older model or some foreign makes, a positive ground may be encountered.

Of course this should be ascertained before installation is even thought of, and if a new unit is being purchased, one with a positive ground sought. If though this was not possible and the installer is faced with the problem of incompatible polarities, the day can be saved by the use of a *polarity converter*. This is a device which turns the d.c. from the battery to a.c. by means of a transistor oscillator, then isolates the supply through a transformer before turning it back to d.c. by rectification, with the opposite polarity. There are also similar devices which use a step-up transformer in the a.c stage so that a 12 V output can be obtained from a 6 V supply. They are useful for driving a conventional 12 V radio or player in a vehicle using a 6 V battery.

The power lead from the unit is in two sections linked by a bayonet-type plug and a holder which carries the cartridge fuse. No connection is possible without a fuse being in place, so make sure that one is present and it is of the correct rating specified by the makers. The supply end of the cable should be pushed through one of the grommets in the bulkhead carrying other wiring so that it emerges in the motor compartment. There it must be connected into the junction box which supplies the car electrical system. Spare connections are usually to be found here for the radio or other electrical accessories.

As a rule there are two connection points available, one going directly to the battery and the other switched by the ignition key. Radio makers often recommend the use of the switched connection so that the radio is dead when the ignition is turned off. The reason for this is that it cannot then be left on accidentally when leaving the car, so running down the battery. Such an arrangement makes listening impossible when parked with the ignition off, and there may well be many occasions when this may be required. In these situations the ignition would have to be left on even if the motor was not running and thus would produce a drain on the battery. So, there is the choice of the two, but as one would have to be very careless to leave a radio or tape player running when absent, it is probably better to make the connection direct.

The switch-on

Now comes the 'moment of truth', the switch on. Hopefully, everything works, and in the case of a radio, the antenna trimmer will have to be tuned to give maximum volume when receiving a weak station at the high frequency end of the scale. Do not adjust on a strong station, otherwise the receiver automatic gain control will start operating and will make accurate adjustment difficult.

If the car ignition circuits have not been suppressed, this will now have to be done as outlined in the previous chapter, but it will be worth while to start the motor to see just how bad the interference is and so give some idea of the extent of the measures required. Try all accessories such as wipers and heater motor to see if they generate interference and therefore need suppression.

11. Trouble shooting

We can divide the troubles likely to be encountered into two categories, those arising from faults in the installation, and those due to breakdown in the actual units. Unless we happen to be hot on radio and electronics, it is not advisable to get involved too deeply in the units themselves, as this can do more harm than good. In such cases it is best to take the faulty unit to a competent repair shop. However, most of the troubles that crop up are usually in the installation and so can be handled by the owner; we will take a look at some of these.

Radio and tape player installations

If a unit has just been installed in a new vehicle and is completely dead, first check the polarity of the car and that of the unit. This point was mentioned in the last chapter, but a considerable number of units are fitted to vehicles of the wrong polarity and it is worth making sure about this before looking elsewhere for the trouble. If indeed the polarity is wrong the most likely result will be a small puff of smoke and smell of burning when switching on! This may not be as disastrous as it seems. While the output transistors may have blown, the most likely effect is the burning of the transistor emitter resistors and the blowing of the fuse. Try a new fuse (after correcting the polarity switch) and the unit may work if the resistors are not badly burnt. If it doesn't the resistors will have to be replaced at the repair shop.

If the polarity is correct and the unit is dead, check that power is reaching it from the battery. If the unit has a dial lamp or power indicator, this will show whether it is receiving current. If not, check the fuse (new fuses have been known to be dud), and whether the fuse is making good contact with both ends of the holder. Should this all prove to be in order, examine the connection in the junction box and the fuses therein. A circuit tester consisting of a voltmeter or a 12 V lamp with a pair of wires will be useful for checking these points. One wire should be clipped to the car body at some convenient point, and the other touched on the point being tested.

With a tape player, the motor can usually be heard running if the ear is held close to it and the surroundings are quiet. This obviously proves that power is getting to it. Alternatively a tape set to the start of the cassette or cartridge can be inserted, and withdrawn after a few seconds. If it is still at the start, then the motor and drive system have not been functioning.

Loudspeakers

If power is reaching the radio or player but there is still no sound, we have to ascertain whether the speaker circuits are working. The first test is to listen with an ear close to the speaker and gradually turn up the volume. There should be an audible hiss which shows that the speaker is in fact in order. If not, turn the volume back quickly as a fault in the speaker circuit could damage the output transistors if they are operated at high volume.

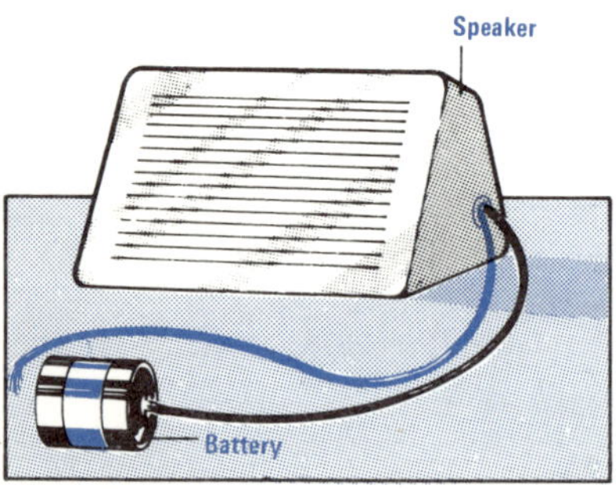

A torch cell touched momentarily across loudspeaker wiring should produce crackles. Silence indicates either speaker or wiring is faulty.

In the absence of hiss or any noise at all, it is most likely that the loudspeaker circuit is in fact the culprit. Switch off the unit and disconnect the speaker; now momentarily touch the speaker wires across a 1.5 V battery (do NOT use the car battery for this test). This should produce a crackling noise from the speaker. If it does not, there is a fault in the speaker or its wiring. Connect the flashlight battery directly across the speaker terminals for an instant, and if a crackle is now produced, there must be a break in the wiring. If there is still no sound, disconnect the wiring from the speaker and try again with the battery cell. Any crackling indicates that there is a short circuit in the wiring, and the speaker is all right, but if it is still dead, the speaker itself is defunct.

With a stereo outfit, it is unlikely that both speakers and their wiring should have a fault, so if there is no sound at all the trouble will probably be elsewhere. The same applies with other symptoms such as distortion or low volume. If though one channel only is affected and the other is normal, a quick check on the speaker can be made by simply changing over the speaker connections. If the fault switches to the other speaker, then it is the player that is at fault, but if the trouble stays with the same speaker, the speaker or its wiring is to blame.

When the speakers are found to be in order and power is reaching the unit, the defect must be in the unit itself, and in most cases this means dismantling and taking it to a repair shop.

Antennas

A common cause of trouble with radios is the antenna. If stations are weak and there is a high background hiss, possibly accompanied by ignition interference, it is almost certain to be an antenna fault. Firstly, check the continuity from the highest rod to the tip of the antenna radio plug. Ideally this should be done with an ohmmeter, which measures resistance directly in ohms, but an indication can be obtained with the aid of a lamp and a battery. Connect the lamp circuit so that the antenna through to the plug is in series with the lamp and battery. The lamp should light at full brilliance. Now connect the lamp directly across the battery and compare the brightness; there should be no difference. If it is dimmer when the antenna is in circuit, there is some series resistance which may be the cause of the trouble. Try it from the lowest section of the antenna to the plug. If the brilliance is now normal, the resistance is between the telescopic sections, a quite common fault. To cure this clean off the rods with a slightly abrasive cleaner, paying particular attention to any rust or corrosion spots, then smear with thin oil.

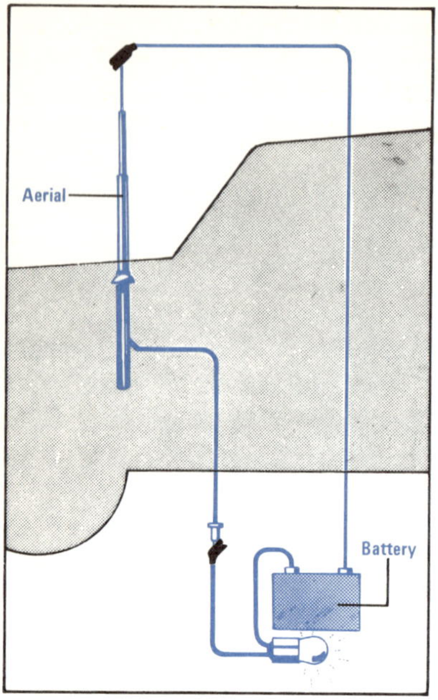

A circuit tester consisting of a dry battery and lamp can be used to test antenna and cable continuity. Clip wires to plug and top section of antenna. Lamp should be no less bright than when connected directly across the battery.

Another antenna trouble is leakage between the antenna and the car body caused by damp mud collecting across the insulator at the bottom where the cable is connected to it. The symptom here is loss of strength and high noise which get worse during and just after a spell of wet weather. When it is dry, reception returns to normal. The remedy is to clean off the mud, removing in particular any that might have got inside the insulator to the connections. A treatment with a moisture repellant may help to inhibit future trouble.

Interference

Interference problems were dealt with in Chapter 9, but there is a fault condition which will not respond to simply fitting extra suppressors. This usually arises with a car that has had a radio installed for some while. The trouble is an increase in interference where at one time there was none. A common cause of this is a poor ground connection to the braiding of the antenna cable. Grounding is accomplished by the clamping component underneath the surface on which the antenna is mounted. This should make good contact with the metal-work, but may not if the securing nut has become loose or dirt or corrosion has crept underneath.

A test can be made by using the lamp and battery device or some other circuit tester. Unplug the antenna plug from the radio and test for continuity from the outer sleeve of the plug to the car bodywork. There must be zero resistance as shown by undiminished brightness of the lamp as compared with a direct connection across the flashlight battery. Even a low resistance may give ineffective shielding and result in interference.

Connections

Loose connections are an ever present possibility with mobile equipment. The effect will depend on what part of the circuit is affected. Trouble of this description in the power line, including the connection in the junction box, loose contact of the fuse in its holder, or a loose connection between the unit and ground due to perhaps a loose screw or bolt securing the rear support strip, will cause crackling and momentary cessation of sound in most cases. The trouble will usually be present when the vehicle is in motion but not always when stationary, as the connection will be affected by vibration. A clue as to the cause can be obtained by watching the indicator or dial lamp, which will flicker if there is a bad power connection. Don't watch it too intently when driving though!

If the lamp remains steady then the crackling and signal loss must be due to a bad joint elsewhere. With a radio, the most likely cause is the antenna or its connections. To test this, a short length of wire could be plugged into the radio socket in place of the antenna plug, and run up over the fascia under the windscreen or out through the window. A short drive should indicate whether the trouble is still present. To minimise the interference which will be picked up without the antenna cable shielding, tune to a strong station. If the fault is now clear, the trouble is obviously in the antenna.

Loudspeaker wiring may also suffer from poor connections, but this does not usually produce crackling, only momentary loss of sound. If the system is stereo, the speaker connections can be changed around to see if the same speaker gives the symptom. If it changes to the other one, then the trouble is in the unit.

Tapping the radio or player sometimes helps to identify the source of the fault. If there is a loose joint in the unit, the tapping will produce the noise. Not all spurious noises arise from bad connections, though. Some can be due to component breakdown in the amplifying circuit and this needs expert attention. However, if the trouble can be positively traced to the unit by the tests here described, at least you can dismantle it and take it to be repaired yourself instead of driving the car to the repairer for testing and dismantling. This should save considerably in repair costs.

Tape player fault diagnosis

Although electronics may be a closed book to most folk, there are some defects in a tape player which can be sorted out. They are those that are mechanical in nature. A very common trouble is loss of treble in the reproduction resulting in a muffled sound, as though the tone control was fully turned down. In nearly every case this is due to a dirty playback head. Oxide powder from the tape in time builds up a deposit across the magnetic gap on the face of the head preventing intimate contact between the head and the tape and also forming a magnetic short circuit. With car tape players the head is unfortunately not so accessible as with ordinary recorders, so the unit must be dismantled from the car and the case removed. Then the head face can be cleaned with some cotton wool on a stick and a liquid head-cleaner or methylated spirit. After cleaning, dry and polish with clean cotton wool.

To prevent deposits forming and needing this treatment, it is good practice to run a head-cleaning tape through the machine once in a while. The tapes are available in both cassette and cartridge form and carry instructions as to how often they should be used. This though should be *before* any deterioration of the sound becomes audible. A cleaning tape will not shift a well established formation of oxide, where hand cleaning is the only remedy.

After a considerable period of use, the head will gradually wear, which produces the same symptom as an oxide deposit. In some cases it can apparently be cured by cleaning, but it will need cleaning again very soon afterwards. This is because a cavity has formed in the head face due to wear, which collects and traps oxide particles much sooner than a smooth surface would. The only answer to this is to replace the head. Although not a big job it is best done by a qualified technician as the adjustments are critical and the soldering rather fussy.

Misplaced or ill-adjusted heads can produce various symptoms. To give the maximum treble response the head must be perfectly vertical. Should it incline from the perpendicular, treble will be lost, in amounts proportional to the angle of deviation. There is an adjustment on the head mounting known as the *azimuth setting*. This is a screw or pair of screws on one or both sides of the head base. The action of these is to rock the head from one side to the other as they are turned.

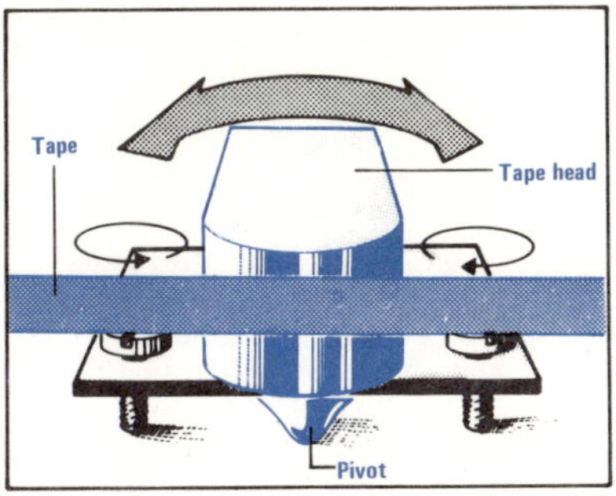

Head azimuth is adjusted by simultaneously turning the two side screws in opposite directions, thereby rocking the head over a centre pivot. Sometimes one screw is fixed and that side of the head support is sprung. Adjustment is made for maximum high frequency response.

If then there is loss of treble, and the head is perfectly clean with little sign of wear, the azimuth may need adjustment. In some cases a worn head will perform somewhat better if the azimuth is slightly altered. A special test tape is really needed for this adjustment, but if you have a cassette recorder in good condition with its heads correctly aligned, a test tape can be recorded on it. This can consist of some prolonged applause, or some keys jangling. When playing this tape on the car player, the head azimuth should be adjusted to give the clearest sound. It will be found that at one position of the screw, the sound is sharp and clear, while only a fraction of a turn either way will reduce the clarity. Care and attentive listening are therefore required for this job. Once correctly set, a blob of paint on the screw will lock it and prevent it turning with vibration.

In fact azimuth adjustment is rarely required with cassette machines, but may be more often needed with cartridge players because of the physical movement of the head with each track change. Some cartridge players have this adjustment brought out to the control panel as a user control. In this case it should be adjusted on a music tape to give the brightest musical sound.

Another trouble with cartridge players is the vertical misalignment of the head relative to the tracks on the tape. This imperfect registration gives reduced signal and increase in background hiss. In bad cases the head gap will encroach on a neighbouring track resulting in crosstalk or the hearing of two tracks at once. To correct this the head must be moved vertically up or down until the interfering track is lost and the sound is at a maximum with minimum hiss. It is common practice for this adjustment to be labelled *fine tuning* and brought out to the control panel. Vertical misplacement of a cassette player head is very unlikely unless by physical damage.

Tape tangles

One problem that seems to dog cassettes is that of tape tangling. The symptom is that the player suddenly stops working, and on withdrawing the cassette quantities of crumpled tape are found to be loose in the playing compartment of the machine.

What happens is that for some reason the take-up hub stops turning so that tape is pulled off the other hub by the drive capstan, and not being taken up, just spills out everywhere. It sometimes wraps itself around the drive capstan, making ejection of the cassette very difficult. The thinner long playing tapes seem more prone to this trouble than the shorter ones.

As we saw from Chapter 5, the take-up hub is not driven directly but through a friction drive consisting of a plate and felt ring. This is so that its speed can gradually decrease as the hub fills up. It follows from this that the torque applied to the take-up hub must be low – it can in fact easily be stopped with the fingers while the machine is running. If then any excessive friction appears associated with the tape being taken up, the hub will simply stop. This friction can be due to rough mouldings of the hub plastic, burrs on the case or deformation of either causing the hub to bind against the casing.

Another possible friction source is between the sides of the pile of tape on the hub and the casing. Low-friction liners (A) are usually fitted to the inside surface of the case to minimise it, but it still can occur. One cause of this is when the tape is stopped part-way through, wound back a short distance and played on. This may be done when searching for a particular passage on the tape, but it very likely will produce a stepped tape edge, with one section not lying exactly on top of the one below (B). This virtually increases the width of the tape pile and so brings extra pressure to bear on the side liners and thus extra friction.

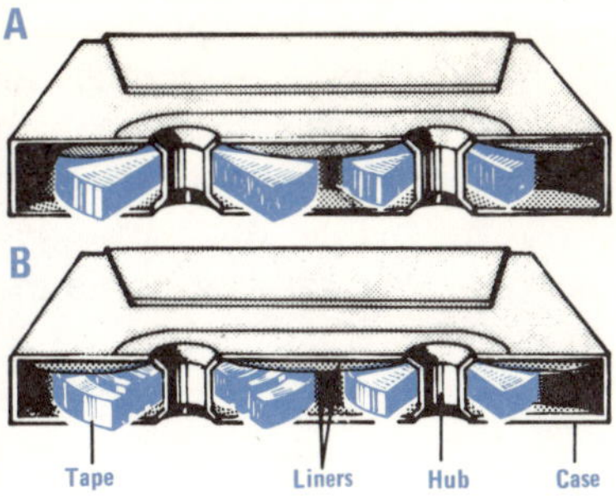

A Tape pile in cassette supported at the sides by low-friction liners.
B If tape pile is stepped by rewinding and fast forwarding over part of its length, effective width is greater and side pressure, hence friction, on liners is greater. This may stop it revolving, causing a tape tangle.

The situation is worsened if in searching for the passage, fast-forward is used as well as rewind, as would be done if the tape was rewound too far. Then not only will the tape become stepped, possibly in two directions, but the fast-forward section will be of a different tension to the others.

So, although tape location facilities are offered on tape players and recorders, and although nothing untoward happens most times they are made use of, the risk of a tangle is increased. When playing cassettes in the car, one usually listens right through the tape rather than pick out certain parts, so the problem is less likely than when playing cassettes at home.

Belts and idlers

Slow or erratic running is usually due to a stretched or perished drive belt, or hardened rubber idler wheel. The belt can be checked by pulling it slightly and letting go. It should return smartly to its former position. If it does so slowly or even not at all it needs replacing, as its elasticity has gone. There should be plenty of tension and no sign of slackness. A hardened rubber drive wheel can also give this sort of trouble. The best way of testing it is to press a thumbnail into the rubber; it should 'give' but return without any nail mark when released.

A regular thumping noise is often caused by an idler wheel, or by an indentation in the perimeter of the rubber roller that presses the tape against the drive spindle. The reason for this is the play button being left in the 'on' position with the unit switched off. In this condition the rubber wheel or roller is pressing against the spindle at one point and so an indentation forms. If the rubber is supple and in good condition, the dent will gradually disappear after an hour or two's running, and with it the thumping noise. With most cartridge and many cassette players the mechanism only engages when a tape is inserted and the machine starts to play; they cannot therefore be left in the engaged position without running. Not all do this, though, and in particular the combined radio/players can be switched off or switched to radio with a tape in the playing compartment. It is wise to make it a rule never to leave a cassette or cartridge in a player when it is not in use.

Wow

Speed variations of a regular nature, termed *wow*, are due to some rotating component. It may be eccentric, have a stiff bearing or be fouling something, thus causing a varying load. To locate the trouble, get the player working with the cover off and the works visible, then try to match the speed of the wow variations with the rotation of one of the parts. The one that matches is the source. A closer examination will then reveal the reason for it.

A common cause of wow is an oxide deposit on the drive capstan or pinch roller. It will be remembered that each cartridge has its own pinch roller, so any trouble due to this cause will be present only with one cartridge. Frictions and other troubles can develop in both cartridges and cassettes, so it is always as well to try another tape before starting to look for a fault in the player.

Oil is one of the worst enemies of a tape player. It will perish rubber parts and get on to friction drive surfaces to cause loss of traction. Very little lubricant is required, motor bearings being of the oil-sealed type, as usually are the flywheel top bearings. Nylon is used extensively for bearings and moving parts and requires no lubrication. An occasional spot may be administered to a pivot or linkage that shows need of it, but it should be kept to the minimum. Should there be signs of oily deposits on the drive surfaces they can be cleaned off with spirit.

As with most items of audio equipment attention to simple maintenance points such as regular head cleaning, and care in operation and handling, should result in a long trouble-free life and many miles of pleasant mobile entertainment.

Index